Analytics and Modern Warfare

Previous Publications

Economics and Modern Warfare (2012)
Psychology and Modern Warfare (2013)
101 Things Everyone Needs to Know About the Global Economy (2012)
Corporate Finance for Dummies (2012)

Analytics and Modern Warfare
Dominance by the Numbers

Michael Taillard

ANALYTICS AND MODERN WARFARE
Copyright © Michael Taillard, 2014.

All rights reserved.

First published in 2014 by
PALGRAVE MACMILLAN®
in the United States—a division of St. Martin's Press LLC,
175 Fifth Avenue, New York, NY 10010.

Where this book is distributed in the UK, Europe and the rest of the world,
this is by Palgrave Macmillan, a division of Macmillan Publishers Limited,
registered in England, company number 785998, of Houndmills,
Basingstoke, Hampshire RG21 6XS.

Palgrave Macmillan is the global academic imprint of the above companies
and has companies and representatives throughout the world.

Palgrave® and Macmillan® are registered trademarks in the United States,
the United Kingdom, Europe and other countries.

ISBN: 978–1–137–39563–4

Library of Congress Cataloging-in-Publication Data

Taillard, Michael, 1982–
    Analytics and modern warfare : dominance by the numbers /
Michael Taillard.
        pages cm
    Includes bibliographical references and index.
    ISBN 978–1–137–39563–4 (hardback)
    1. Military art and science—Forecasting. 2. Mathematical statistics.
    I. Title.

U163.T35 2014
355.0201'51—dc23   2014024357

A catalogue record of the book is available from the British Library.

Design by Newgen Knowledge Works (P) Ltd., Chennai, India.

First edition: December 2014

10 9 8 7 6 5 4 3 2 1

Transferred to Digital Printing in 2015

Contents

List of Figures and Tables   vii
Preface   ix
Acknowledgments   xiii

Introduction   1

Part I   Descriptive Analytics

Chapter 1	Descriptive Statistics	11
Chapter 2	Modeling	23
Chapter 3	Comparative Assessments	37
Chapter 4	Data Diagnostics	47
Chapter 5	Challenges and Limitations	61
Chapter 6	Suggestions for Future Research	65

Part II   Predictive Analytics

Chapter 7	Probability Modeling	73
Chapter 8	Correlative and Regression Analyses	83
Chapter 9	Geospatial Intelligence	93
Chapter 10	Challenges and Limitations	103
Chapter 11	Suggestions for Future Research	107

Part III Operational Analytics

Chapter 12	Quality Assessments	115
Chapter 13	Efficiency Analyses	125
Chapter 14	Risk Management	139
Chapter 15	Asset Management	153
Chapter 16	Challenges and Limitations	171
Chapter 17	Suggestions for Future Research	173

Afterword	177
Critique of Current Methods	183
Index	189

Figures and Tables

Figures

1.1	Bell curve	15
1.2	Box plot	19
2.1	Curvilinear scatterplot	34
4.1	Q–Q plot	51
7.1	Venn diagram	75
12.1	Six sigma distribution	117
12.2	Quality goal posts	118
14.1	Composite risk management	144
14.2	Operations portfolio	150
15.1	Decision tree	169

Tables

3.1	Reference guide	44
4.1	Threats to external validity	58
4.2	Threats to internal validity	58
12.1	Six sigma statistical analysis	117
13.1	Productivity potential without cooperation	130
13.2	Productivity potential with cooperation	130

Preface

It is important to state from the very beginning that this is not a math book, but rather, a book about the way in which math can be used to create strategy—specifically, statistics. Far too often people are confused or intimidated by analytics, and just as frequently people also tend to believe they will never use analytics in their own career. This book is written specifically to address these issues—rather than simply providing the exercises and tiny details of obscure mathematical rules, the focus of this book will be on explaining what one is capable of accomplishing using analytics. Emphasis will be given to what one can expect when utilizing these analytics, the role they currently play, the role they are capable of playing, and real examples of how they have been a decisive factor in important operational and strategic hurdles. Also provided will be simple descriptions of what is happening in each analysis, broken-down into series of easy-to-follow steps so that they can be repeated without necessarily having any prerequisite knowledge in mathematics. The bulk of the book, though, is focused on written descriptions of when to use these calculations practically, what they are capable of accomplishing when utilized properly, and what to expect as a result of their use. It is not the goal of this book to teach you mathematics, yet a prerequisite knowledge of mathematics is not required to understand this book, either. It is the goal of the book to teach you how mathematics can be incorporated into organizational operations and strategy in order to create a profound competitive advantage, and to encourage the reader to utilize them more greatly in operations, either by utilizing their existing knowledge of mathematics, or by tapping into talent that can provide detailed analytics. To accomplish this, one must understand which tools are available, how each is used, and what results from their use. You do not need to understand the design of a tool in order to use it, and so this book will teach you the use of these tools rather than their designs.

This book is written in such a way that the function of each type of analytic is isolated and described. Each chapter is dedicated to a single type of analysis that can be performed, and the reasons one might want to perform it. The reason for the focus of their potential application to warfare is simple: military goals are among the most difficult to accomplish, given their often extreme nature, and their success or failure is among the most easily recognizable, since the results of military actions are rarely encountered under normal circumstances allowing one to "reject the null hypothesis," as will be discussed throughout this book. It is true that the topics throughout this book have vast potential that is not currently being met by military forces, and that will be the focus of this book, but this is largely intended to demonstrate the current limits of what is possible, to provide perspective on just how far even the private sector is from achieving what is possible, even now.

You are about to pursue a path that has been pursued before, and learn from these historical accounts to derive from them the foundations of a recent revolution in the use of quantitative analytics in everything from business intelligence, law enforcement operations, and military strategy. Nothing described throughout this book is new or cutting edge—everything already proven but is currently not being applied in a manner in which to realize its full potential, allowing for expanded application, and the ability to derive fundamentals of new applications upon which more advanced groundbreaking research can be built.

Without knowledge of the world around us, we are simply guessing at what is, what may be, and what we should do. Analytics is nothing more than a way of collecting measurements and calculating answers to questions to generate the knowledge we need to generate the knowledge we need to properly function. It is the differences in this knowledge that will define who better understands the powers at work, who can better predict the outcome of events, and who can better operate within any environment, much less one of such great intensity as a battlefield.

Acknowledgments

This book is dedicated to the National Organization for the Reform of Marijuana Laws (NORML). The analytics are clear: a large percentage of the conflict which exists in our world would be eliminated with legalization. As the global narcotics trade is among the largest markets in the world, half of which is composed of marijuana, and prohibitionist policy has placed it directly in the hands of those who would use the profits to fund violent crime, legalization would decimate the active presence of street gangs, drug cartels, terror cells, mobs, and a variety of other organizations. The elimination of overcrowding in prisons would stop violent offenders from being released early, and keep non-violent offenders from being exposed to, or recruited by violent ones. Ending the long-term impacts of criminal drug offenses on people seeking employment or education will also help to prevent them from turning to criminal activities out of disgruntled desperation. The reallocation of law enforcement resources from enforcing prohibition to preventing and stopping crimes of violence or property damage would increase the effectiveness of existing resources, while tax revenues raised could be used to fund prevention measures. As the world continues to look for ways to fight the monster that is global violence, one of the most effective methods is to simply starve the beast, and there is a wealth of data to show it. The implications of marijuana legalization reach further than most people realize, and the work that NORML does to make legalization a reality is among the most important global policy efforts being made today. I urge the world to support the efforts being made by NORML as an issue of public health and safety, and a very important tool of strengthening the economy. The data analytics illustrate clearly that this is necessary.

Introduction

Analytics tend to look pretty scary, and the thought of deciphering a book of mathematical principles with which you are not familiar can be extremely intimidating. Some people claim that they would prefer to face an entire army of insurgents than a single statistics examination, but if you are a big sports fan, then odds are good you have used applied statistical analytics already! For example, if you watch baseball it is likely that you have followed the performance of a particular player, tracking the number of games in which they have played, the number of times per game they successfully hit the ball or reach a particular base, the number of times per game they scored a point, hit a homerun, stolen base, and so forth. You may have even used the averages between these games to estimate what their performance will be in a future game, and try to predict which team will win a game or make it to the championship. You may have even applied this information to infer what a team's strengths are, and what they need to work on in order to be competitive during the rest of the season. Casinos are also an extremely common place where people frequently utilized applied analytics, in the form of probability theory. Calculating the odds of an outcome on any game is relatively simple once you know the basics, and that will help you improve your successes by determining the amount of "house advantage" (the degree to which a casino maintains a higher probability of winning than the player) inherent in each game, how to make individual bets, and how to develop multi-bet strategies. These examples are applications of statistical analytics, which are used for fun by a large percentage of the global population, many of whom would claim that they hate math when they encounter the same analytics outside of the context to which they have become accustomed. Statistical analytics can actually be easy to understand when presented in the right way. Even researchers utilizing the most complex of formulas and quantitative models will present their work to others using simplified explanations with lots of pictures and graphs.

It has been said that mathematics is the language of the universe, able to define and describe all things. Statistics, as a specific field of mathematics, includes the development of tools that allow for one to identify truth by collecting measurements and analyzing the resulting data. These analytics yield meaningful information that communicate the nature of those things being measured, as well as methods in which to mathematically measure uncertainty, dedicated to determining whether something is true or not, or the likelihood of making a particular observation. Statistical analytics are used to evaluate measurements of all things to understand them, to predict what the future holds by calculating patterns and relationships that exist, develop models that represent processes and operations so that their interactions and functions can be accurately assessed and managed, and to make informed decisions that will have outcomes known to be optimal before they are made. When combined with even a basic understanding of strategy, the use of analytics creates a significant advantage and has the potential to become a core competency, in itself. Whoever best implements quantitative analytics into their operations, and develops the most accurate models, will have better information and more effective operations.

The use of analytics has vastly increased in popularity in recent years in a number of different applications. Law enforcement has made great strides in reducing crime through preventative analytics, which has come to be known as predictive policing. In the private sector, business intelligence is a fast growing industry, as a variety of companies of varying size exclusively dedicate themselves to offering services in research, data management, and analytics to client businesses. Though the services they offer are not new ones, by any measure, businesses across most industries have been largely unaware of the potential that analytics hold for improving operations, developing market strategy and competitive positioning, and for identifying and managing those external factors that influence the success of the company. There are broad investments being made in the "big data" industry—into data research and analytics firms—throughout this world, many regions attempting to attract these investments, such as attempts by the US Midwest to brand itself "Silicon Prairie," and the advances being made in this industry are benefitting not just private companies, but also defense and intelligence operations, as well.

The use of qualitative analytics in strategy is quite ancient—and even the use of quantitative analysis has seen varying degrees of popularity, but most of the precise mathematical tools that are available today are relatively new, with their origins dating back only to about the seventeenth century, around the time of the Thirty Years War, though the two were not related. WWII was very much a renaissance for defense and intelligence industries, as innovation had reached its peak, new methods were being explored to end what had become the largest war in all history, and more than in any time in history scientists were being tapped to develop new tactics and weapons to gain a strategic advantage. It was during WWII that statistical analytics had become popularized; modern quality management techniques such as Six Sigma have their roots in analytics developed during WWII to improve operations success and resource utilization. Predictive analytics utilizing a variety of variables ranging from resource market pricing, to geographic movements, to just about any other data that could be collected were used to generate estimates of where strategic assets are being held, what plans the opposition has, and what the outcome will be of particular conflicts. A great degree of progress into the utilization of analytics was used during this time period, but they were largely discarded after the war ended in favor of more traditional, qualitative type of analytics. The use of espionage and spies became extremely popular during the Cold War, as did the direct observation via satellite, as the US Department of Defense announced that they were pursuing a satellite intelligence system in 1948, one year after the start of the Cold War. It was not until the post-2001 era when many of these analytics were rediscovered, as steps were taken by the US

Federal Government to use expanded mass-surveillance methods to collect huge volumes of data from internet and telephone sources. The vast volume of this data becomes nearly meaningless—impossible to decipher—without the use of quantitative analytics intended to identify specific key data types and trends. Despite the increasing emphasis placed on quantitative analytics in the intelligence community, the defense industry still fails to keep-up, particularly in the realm of those operational improvements that have been realized in the private sector.

In the world of analytics, there are three broad types of indices: lagging, leading, and coincident. Lagging indicators are those that provide insight of what has happened, what is currently happening, and what will happen, by utilizing historical data—they are highly accurate but, given their historical nature, tend to be a little slow to be helpful. Since lagging indicators are those that tend to change after the event being measured does, they are used frequently for confirmation, or in watching for trends over time to predict what is currently happening or will happen in the future. Leading indicators are those that change before the event being measured does, so that they provide insight into future events, with varying degrees of accuracy. Coincident indicators are those that occur at the same time as the event being measured. In all cases, it is important to remember that indices do just one thing: indicate. All analytics provide insight—highly valuable information that can be used to make decisions, improve operations, predict future events, develop strategy, and so forth—but that is all. The information derived from analytics must be utilized in context, and with careful application.

This book is divided into three major parts. Part I discusses descriptive analytics, which are ways of measuring the world around us. Descriptive analytics define for us what is fact, whereas simple observation is highly flawed, and allows us to identify specific important traits in the internal and external environments critical to the development of strategy. Part II discusses predictive analytics, which allow us to accurately determine what is going to happen in the future, or what the outcome of a given decision will be. This allows us to make proper decisions, prepare for future events, and understand the role of current events in the things to come so that they can be manipulated and the future altered. Part III discusses operational analytics, which are equations and models developed to better understand the functions and processes within an organization, its role within the greater context of achieving its goals in a given environmental setting, and how resources can be managed in a way to accomplish those goals to their maximum potential as fast and efficiently as possible, without error or risk. An afterword discusses some special-use analytics, which apply to psychological and economic warfare, known as psychometrics and econometrics, respectively, as well as the importance of computer and information warfare.

PART I

Descriptive Analytics

Remaining strategically competitive can be difficult enough under the best of conditions, but unless one can accurately describe what exactly is happening with precision, it can be almost impossible. Most people think they understand what they see, but the limitations of our senses are severe and profound, making personal observation nearly useless. Confidence without proper analysis comes not from an understanding of what has happened, but from an inability to conceive of other possibilities. For example, during reconnaissance should a person observe several convoys leaving an opposition stronghold they may come to the conclusion that the location has reduced its operating strength as a result of having fewer people and fewer resources readily available. Should the observer be highly confident of this, it does not mean they are right, only that they have not considered the possibility that the movements witnessed may not significantly decrease functional strength at all, or that it is only a short exercise, or that it is only one element in a larger movement that includes incoming resources as well, or that the convoys were only moving a short distance to expand their regional operations, or any of a number of other possibilities. Casual observations about the movement of resources lead to common misperceptions during peacetime, as well, when the people of a nation believe that jobs or operations are being sent overseas when, in fact, national production continues to reach record highs and operations move from highly populated, high-cost areas to cheap, rural areas where fewer are there to witness the change. During peacetime, these types of sloppy conclusions drawn from observation are, at worst, an obnoxious political talking point, which begets little in the way of action, but during a conflict this can mean the difference between life and death for a number of people—the difference between winning and losing a battle. When there is no room for error, then one's actions must be mathematically exact—the same unforgiving precision that is the bane of so many children during their math lessons must now become the bane of the opposition.

There are two broad categories of statistical analysis: descriptive and inferential. This can be a bit counterintuitive, because both are used descriptively, but descriptive statistics describes the thing being studied while inferential statistics uses descriptions of the data collected to infer descriptions of the thing being studied. Essentially, descriptive statistics provide information about the traits being measured while inferential statistics are used to determine whether the descriptions are valid, and the strength of the data in explaining or predicting some variable or relationship between variables. When testing data, the idea being tested is known as a hypothesis, which is a specific statement about the thing being studied that is being assessed for accuracy. If the results of the test are strong enough, they are said to be statistically significant, which means that the probability of the test results being caused by chance are so small that the very idea is rejected. This is known as rejecting the null hypothesis. The hypothesis for which you are testing is sometimes called the alternative hypothesis, while the null hypothesis states simply that the data being tested is not conclusive. It is a common mistake to say that the null hypothesis states that the alternative hypothesis is wrong but, in reality, you do not reject the alternative hypothesis, you are only incapable of rejecting the null hypothesis. When this happens, the possibility still exists that any significance in the data is merely the result of chance when doing sampling (discussed more in chapter 2). In order to reject the null hypothesis, the probability of chance being a factor must be very small, the most commonly chosen significance levels being 5 percent, 1 percent, 0.1 percent, and 0.00001 percent, depending on application; these are the milestones of probability, which must be surpassed—in other words, there must be less than 5 percent probability that the data collected were the result of chance rather than a systematic relationship.

When the results of a test are wrong, it falls into one of two types of errors. Type I errors are commonly known as false positives, which means the analysis shows something significant in the data when, in fact, there is nothing in reality. Formally, a type I error leads to the incorrect rejection of the null hypothesis. Type II errors, in contrast, are commonly known as false negatives, which means that the analysis shows that there is nothing of significance in the data when, in fact, there is something in reality. Formally, a type II error is the incorrect failure to reject the null hypothesis. It is in human nature that people are extremely prone to making type I errors. From the perspective of evolutionary psychology, the people who were prone to making type I errors would observe sounds and run away, thinking it was a predator rather than potential prey. They may lose a squirrel breakfast, but they will be alive to pass-on their genes. In contrast, those who are prone to making type II errors may observe sounds and think it is nothing significant when, in reality, they are about to be attacked by a predator. Over the millennia, this selective survival of specific psychological tendencies has created a human race extremely prone to finding patterns, which simply do not exist; they see faces in the moon or in a piece of toast, they form superstitions such as the belief that some action they took led to a particular outcome which then becomes a ritual such as before sporting events, they attribute particular outcomes or actions to the metaphysical despite a clearly explainable cause, and they make relationships between things when there simply isn't one. Such errors are also possible when performing analysis, but the level of significance we place on the results of an analysis is critically important in evaluating the likelihood of such as error being the case. For example, the standard for announcing the discovery of a new particle in physics is 99.9999 percent certain, leaving only the very tiny probability of the data being the result of a type I error. In using this analytical approach, not only can we be extremely confident in what we are observing, but we can also be extremely precise in the degree to which we are uncertain, and assess the sources of the unexplained portion of the equation (discussed more greatly in chapter 2).

In the end, speculation, opinion, and observation—these are just different ways of stating the same thing: guessing; while measured and analyzed data provide facts with which truth can be derived, free of subjective interpretation, and error. Throughout part I readers will learn how to collect data and quantitatively describe it, how to compare the measurements and capabilities of two entities to determine whether there are any significant differences, as well as how to take advantage of those differences by developing models that both visually illustrates the nature of the problem at hand as well as creates standardized equations, which can be utilized repeatedly to find an accurate solution. Descriptive analytics are used extensively in the remainder of the book, as well, since the calculations and data included here are necessary to perform the predictive analytics discussed in part II, and the operational analytics discussed in part III.

CHAPTER 1

Descriptive Statistics

Most people can qualitatively describe what they see within certain limits—events that have occurred, the appearance of things, the direction in which things are moving, and so forth. Unfortunately, the usefulness of these basic observations is extremely limited, incapable of distinguishing between large volumes of things, identify traits that are not readily visible, and influenced by a high degree of subjectivity. Personal observation includes a large amount of estimation and the use of neurological responses that are frequently flawed. Illusions function by tricking the natural processes the brain uses to understand sensory information, and personal interpretations of information of events have a high rate of error resulting from insufficient information and the projection of one's own beliefs and ideas on the external world. When even simple, qualitative analysis exists beyond the realm of our natural observational capabilities, it should then come as no surprise that our ability to accurately analyze anything with any amount of quantitative certainty is completely zero; our ability to measure anything in a manner that is useful in making precise decisions or predict future outcomes is limited to our access to measurement tools, which we do not possess naturally. Since such casual estimation lacks precision and functionality, this makes it little better than guessing. It is for this reason that descriptive statistics are necessary.

Descriptive statistics are some of the simplest yet most fundamental of analytics that one can perform; not only do they provide highly valuable insight into the world around us, but they are also the foundations upon which all other analytics are built. They allow us to understand the nature of things around us in a way that transcends our senses, or even what one might consider logical common sense. Measuring a single item can tell you a good deal about that item, and measuring several more of the same type of item will provide even greater information about the range of possible measurements you will encounter, what measurements are typical, and the kinds of trends you can expect to observe. For example, in measuring a single person, you can measure their physical endurance, their eyesight and how well they can aim, their response time, their knowledge of strategy and tactics, and much more that would tell you about how effective they will be on the battlefield. That one person, though, may or may not be typical, and even if they are, this says nothing of how probable it is to encounter the atypical.

These most basic of analytics describe reality, both the certain and uncertain, using quantitative measures and quantitative descriptions of qualitative variables, so that decisions can be made with certainty rather than by simply guessing. They allow us to take known facts, the data of exact measurements with little meaning independent of context, and use them to derive truths—answers when questions arise from imperfect access to information, or imperfect ability to evaluate and process information through casual observation. This chapter focuses on describing how to perform simple descriptive analytics so that one can do more than provide vague or broad descriptions, instead of allowing them to define with surgical precision the exact nature of the opposition, themselves, and the conflict at hand. This chapter also focuses heavily on introducing the basic mathematical concepts that will be utilized in later chapters to show how they can be used in decision-making, including defining the different types of variables, which can be used to describe, categorize, and measure various elements of the things around us.

In this world of data, more is better, because statistics are used not just to describe those things being measured, but also the data itself in order to derive as much information as possible. Each type of measurement taken is called a variable, which is any value that can vary. There are several types: nominal, ordinal, interval, and ratio. Nominal variables are those that categorize, but which have no definite order. It is extremely common to collect data on national citizenship, type of skill set, organizational affiliation, and other things which, by themselves, have no comparative values. Ordinal values are those which can be placed in some kind of order but which have no quantitative measurements, such as rank. Both nominal and ordinal variables are known at categorical, because they can take-on only a limited number of possible values which categorize data rather than measuring it. By contrast, interval and ratio variables are known as continuous variables, because they can take-on a theoretically limitless number of measured, quantitative values. Interval variables are similar to ordinal variables, except that they have meaningful quantitative values, such as time. 8.00 a.m. and 10.15 a.m. can be categorized quantitatively, and even incorporated into simple comparisons, such has having a 2.25 hour difference between them, but since it is in their nature that their values are derived comparatively as an interval, you could never multiply or divide between them; you could not divide 10.15 a.m. by 8.00 a.m. and get any sort of meaningful value. Ratio variables, however, have an absolute value using 0 as a reference point. Using time, again, as an example, 20 minutes and 5 minutes are each ratio variables; each have meaning derived from its own volume, 5 being ¼ or 25 percent of 20. All data belong to one type of variable, and each variable being measured has a single type of data, though the difference may be distinct. When studying camouflage, color is a nominal variable since there is no way to put into order such values as blue, green, gray, and so on, however, you can measure the amount of light reflected off each color as a ratio variable. In each case, you are measuring the same thing—color—but measuring different variables associated with color. Understanding the types of variables is extremely important not only because the types of statistical tests that are performed will depend on the types of variables being tested, but because understanding the different ways in which the world can be measured, evaluated, and compared, is a decisive factor in the success of an operation. The successes of Napoleon Bonaparte, in many ways, could be attributed to the importance he placed on mathematics and science, and he often associated himself with mathematicians, physicists, and engineers, all of whom play a large role in the success of any military, as demonstrated by the US Army Corps of Engineers and the work they have accomplished, or Albert Einstein and his work in what was, at the time, theoretical physics, which led to the creation of the atomic bomb and shaped global military policy for nearly a century.

There is another way to categorize variables, which are dependent entirely on the relationship each variable has to other variables: by whether each is dependent or independent. A dependent variable is one whose value is shaped by the value of some other variable. How well a person can shoot, as measured by the rate of successes in hitting a target, is dependent on several other variables, such as the volume of time spent practicing, eyesight, and control over extraneous movements in the body and hands (keeping them steady). These variables that determine the value of the dependent variable are known as independent variables because their value is independent of any influence within this context. Each independent variable, though, is also a dependent variable in some other context; eyesight is dependent on genetic variables, environmental variables over the years which can damage the eyes, and corrective variables such as glasses or surgery. As all the variables around us determine outcomes and are, in turn, determined by other variables, by collecting sufficient volumes of the right types of data it becomes possible to "read" reality, understanding past, present, and even future.

Upon collecting data, the total number of measurements that have been collected of a single type is denoted by the letter n. In other words, if measuring 10 distances traveled and the amount of time it takes to make each trip, then 10 would be your *n* value, since there are 10 things being measured, which yield 10 of each type of variable. Simple counting such as this can provide extremely useful information, depending on the context. Heat maps, for example, are common ways to illustrate this sort of data, wherein a geographic map broken into equal-sized units are given a specific color (usually in increasing darkness of a single color) depending on the volume of a particular item within that region, or over time on a timeline—such as population, density of the distribution of specific types of resources, or anything else that relies on simple counting. That is just the start, though.

The word "average" has come to mean, in most people's minds, one specific type of average known as the mean, when there are actually three separate types of averages. Formally, they are known as measures of central tendency, which is very descriptive of what they accomplish: measure the tendency of data to most commonly exhibit a narrow range of values near the center of the total range of values, with a smaller proportion of the total data to have values significantly higher and lower. The distribution of values is most commonly illustrated using something called a bell curve (see figure 1.1).

This is the curve that graphically illustrates the distribution of values of a data set. The higher points on the curve mean that those values are observed in the data more commonly than the values at the lower points in the curve. If you drew vertical bars that grew one unit taller every time you encountered a value, then the tops of those bars would fit perfectly within the curve. Notice that a large percentage of the total data falls within a relatively narrow range of the total range of values. Yes, there are some values that are much higher than the center, and much lower, but the vast majority of the values fall somewhere near the middle. This is typical of data sets. Averages measure the data in a way that finds the middle, each one with pros and cons.

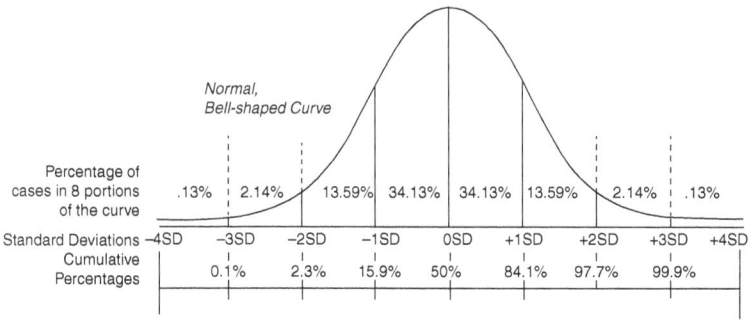

Figure 1.1 Bell curve.

The mean is what is most commonly referred to when people colloquially refer to the average, and is denoted $\bar{x}$ or μ depending on whether one is analyzing a sample or the population, which will be discussed further in chapter 2 on "Modeling," so we will be using the letter m as a generalized reference to the mean. The mean is formally calculated as $\Sigma x/n$, which means that you add-up all the values in the data set, then divide it by the number of values. That first symbol, Σ, is a sigma and indicates that values should be added together; in this case, all the x's, which refers to each of the values in the data set. Each value in the data set is an x, so you add them together, then divide by n, as defined earlier. The mean has the benefit of being extremely accurate if the data are evenly distributed, but it also suffers from problems in that it is easily distorted when the data are not evenly distributed. For example, when there is a value that is extremely high or low, it will cause a significant shift in the mean.

The mean described above is known as a simple mean. There are other types that can be used depending on what you want to know. A moving mean, for example, tracks changes in the mean for a specific range over time. This is common in operational analytics, where one might want to track trends in a 30-day average. A moving mean would start by calculating the mean for the previous 30 days, in this case, and then every day the newest value would be added to the data set and the oldest value dropped from it. You can use a moving average over space, as well, by choosing a specific area and measuring the mean within that area, then moving it one unit in any direction to determine how the average changes. In contrast to a moving mean, a weighted mean places more importance on some values than on others by attributing to each value a percentage. For example, if you want to calculate the average performance of your operations, it might be best to measure the performance of each division, and then weight each by the percentage of total resources allocated to that division. Perhaps, you want to place more emphasis on recent data versus past data without actually dropping data as you would with a moving average, you could distribute a multiplier across the data that more greatly influences data that was collected more recently. Either a moving average or a weighted average can be implemented in any of a variety of ways, depending on what you are attempting to measure. Moving means track trends and patterns in means, and weighted means place more emphases on some data than other data, and they can be utilized in various ways. Both of these, though, will also be subject to the same strengths and weaknesses as a simple mean; a moving mean will be influenced by uneven distributions of data so long as they are present, and even when placing little emphasis on the uneven portion of a data set in a weighted mean, that unevenness will still have an influence, though a smaller one. Means, no matter what type you are using, will always have these problems.

The other two types of averages are not subject to this same type of distortion, the first of which is known as a median. The median is the exact center of a data set; simply line-up the values from smallest to largest, then find the middle value. If there is an even number of values, so that there is no single observed value in the middle, then the median is found by taking the two values in the center and calculating the mean between them. The median has the benefit of being, by definition, in the exact center of the data set, that does not necessarily mean that it will be in the exact center of the values; it is guaranteed that 50 percent of the values will exist above the median, and 50 percent below the median, but the values themselves may cluster at some point in the data set, or there may be extreme values. Therefore, the median and the mean have opposite concerns; the mean is the center of the values but not necessarily the data, while the median is the center of the data but not necessarily the values. The final type of average is called the mode, and this is simply the most commonly observed value in a data set—it is the value that appears most often. Data sets will sometimes have more than one mode, which creates a curve that has two or more peaks, rather than just one. With sufficiently large data sets, these tend to be somewhat rare compared to their one-peaked counterparts, however. A data set is evenly distributed when mean, median, and mode are the same value—it is at this point that the exact center of a data set is clear. When they are not the same, it means that there is some variation in the distribution of the data that must be explored, and the bigger the difference there is between the three types of averages, it is likely that the data set is even more greatly shifted to one side or the other. Before this shift can be understood, though, the distribution of the data must be analyzed.

Measuring the distribution of data provides vital information about the range of possible values, the range of likely values, how widely values are dispersed or clustered, and even provides a method of comparing unlike values. The simplest measure of data distribution is called the range, which measures the smallest interval of values that contain all the data in the set. The range is calculated by starting with the largest value and subtracting the smallest value, which provides information about the entire dispersion of data but says nothing about the density of distributions. A data set may have a total range of 100, only to find out that all the values are between 0 and 10, except one observation of a 100 value, so the range would only show that a the values were dispersed over 100 value units making future observations over that dispersion possible, but would not show that anything higher than 10 was nearly impossible. To measure information about the range of values that are likely to be encountered, rather than those that could possibly be encountered, one must use the interquartile range (IQ range). The IQ range is exactly what the name says: the range of the middle two quarters of the data set. To measure this, divide the data set into four equal parts of equal volumes of x, start with the highest value of the third highest quarter, and then subtract the lowest value of the second highest quarter, as in the following:

With a data set {1, 5, 6, 7}, the middle two quarters are {5, 6}, making the IQ range $6 - 5 = 1$.

A useful tool for illustrating the distribution of data using range, IQ range, and mean is called the box plot, as shown in figure 1.2.

Another way to look at the distribution of data, rather than the range of possible values, is by measuring the degree to which the values deviate from the mean, called the deviation. The deviation of an individual value is measured as $x - m$, and each value in a data set has its own unique deviation, yet the data set itself also has a deviation measured as $\Sigma(x - m)$, which requires you to add together all the individual deviations. The deviation provides information about how much the data are spread from the mean, but this is not very useful without some kind of reference point, since a data set can have a

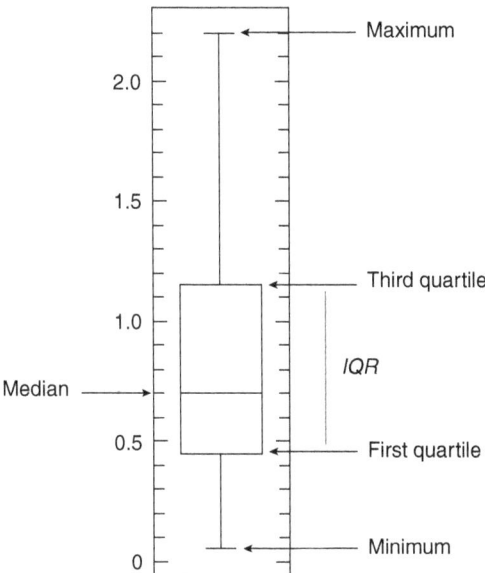

Figure 1.2 Box plot.

very high deviation value even if all the values are very close to the mean if there is simply a very large number of values. That is why the average deviation is more commonly used, which takes the mean of the deviations by dividing the deviation by the number of values, $[\Sigma(x-\bar{x})]/n-1$. This provides useful information about how far from the mean the values exist, on average, making it useful as a way to assess the density of the distribution of data but, as with other forms of means, this can be easily distorted by uneven or extreme distributions. Note that the notation $\bar{x}$ was used instead of m, which is because the average deviation is used only in measurements of population data. Also note that the deviation is divided by $n-1$, rather than just n. This $n-1$ adjustment is necessary for many calculations, which gives us a sample of a population for the analysis of the sample to properly represent the value of the population. If you were measuring the population, rather than a sample, you would divide by n, rather than $n-1$, and use the μ notation for the mean. A slight difference to get the same result, depending on whether you are analyzing the population or a sample.

The average deviation still has problems, however, in that negative values will cancel out calculations of deviations. So, if one value is 10 away from the mean, and another is -10 away from the mean, then the calculation will yield a deviation of 0, indicating that all the values are exactly the same—obviously not correct. To correct for this while maintaining a representative value of the amount of deviation from the mean, a slightly modified version is used, called the variance: $[\Sigma(x-\bar{x})^2]/(n-1)$. It is calculated just like the average deviation, except that each values deviation from the mean is squared. This is one step away from the final type of calculated deviation, which is by far the most commonly used, and not only provides a wealth of valuable information about any given data set, but also allows for the distributions of unlike data sets to be readily compared on an equivalent basis, and it is called the standard deviation. The standard deviation for a sample is calculated as $\sqrt{([\Sigma(x-m)^2]/[n-1])}$. In other words, it is just the square root of the variance, which is done because the variance requires you to square the value deviations, so the inclusion of a square root after the calculation is finished corrects for this. The standard deviation is notated using σ or s, again, depending on whether you are measuring the population or the sample, respectively. The standard deviation provides valuable information about how far from the mean the data are distributed, by establishing a unit of equivalent division. In other words, just as IQ range establishes a unit of value measurement by splitting the data into 4 chunks of equal volume, the standard deviation establishes a unit of distribution measurement that describes in standardized terms the deviation from the mean. Refer back to the curve illustrated in figure 1.1. Below the curve there is a bar that states the mean is found at 0SD, or 0 standard deviation units from the mean—of course, the mean does not deviate from the mean, at all. Then, as you move away from the mean in either direction, you will eventually make it to 1SD from the mean, 2SD from the mean, and so forth, and the number of standard deviations you are away from the mean will determine what percentage of the total values will be included within that range. All the values within 1SD from the mean (meaning 1SD lower and 1SD higher, together) will compose slightly over 68 percent of the total values. 1.96 standard deviations from the mean will include 95 percent of all the values, 2.58 standard deviations from the mean will include 99 percent of all values, and 3.29 standard deviation will include 99.9 percent of all the values in the data set. This is an extremely useful way to measure distribution because it describes distribution in terms of probabilities, which allows for a comparison of unlike data sets. One data set may be working on values between 0 and 1, while another working on values of 1 million and 10 million, which would make using range, deviation, or variance very difficult to make sense of, but by using standard deviation it becomes possible to appropriately compare the distributions. The implications of this will be discussed at great length throughout many of the chapters of this book.

Now that both the measures of central tendency and measures of distribution have been explored, it is possible to define the shape of the data in terms of whether the distribution is even or uneven. There are two terms that describe the shape of the data set, called skew and kurtosis. Skew refers to whether the data are distributed more or less densely on one side of the mean. Take, for example, a data set that has one extremely large value—the curve will have a normal tail on the left and a tail that stretches very far to the right, which will cause the mean to also be higher (further to the right) than it would normally, causing distortion in analysis. Skew does not have to have any extreme values, as long as the values are less densely clustered above the mean (resulting in a positive skew) or below the mean (resulting in a negative skew). A simple estimate of skew can be performed as $(m-v)/\sigma$; start with the mean, subtract the median, then divide the answer by the standard deviation. A data set with no skew will have 0 skew. A set whose mean is higher than the median will have a positive skew, meaning that the center of the values is higher than the center of the data, indicating that the values in top half are more greatly distributed. A set whose mean is lower than the mean will have a negative skew, meaning that the center of the values is lower than the center of the data, indicating that the values in the bottom half are more greatly distributed. Kurtosis refers to how homogenously the data are distributed across the entire data set. Platykurtic data sets ($k<0$) are more homogenous in the distribution of the data, so that no values are observed much more frequently than others, which causes the distribution curve to be shorter and flatter, like a plateau rather than a hill. Leptokurtic data sets ($k > 0$) are very narrowly distributed, with the values around the mean being observed far more commonly than the data observed at higher standard deviations, causing the curve to become very tall and narrow. Mesokurtic data sets ($k = 0$) are those that adhere to something called a normal distribution, causing the bell curve shape shown in figure 1.1.

A normal distribution is any data set that adheres to certain parameters; the mean, median, and mode are all the same value, and they are adjusted so that their value equals 0; the standard deviation is 1; and both skew and kurtosis are 0. A very large ratio of everything that can be measured will result in a normal distribution when the data set of those measurements gets large enough, which is convenient because data that adheres to a normal distribution tends to be more easily analyzed than other types of data. Many types of tests rely on the assumption that data are normally distributed which, when violated, requires an analyst to use other tests that tend to be somewhat less exacting. When data are normally distributed, the natural presumption that most people have about the average of a data set, regardless of which average to which you are referring, is true, as is the presumption that many people have about the distribution of data. The larger your data set gets, the more likely it is to become normally distributed but, even if you find that your data violates this assumption, carefully describing the data using the analytics described throughout this chapter can help to effectively describe whatever it is that has been measured.

CHAPTER 2

Modeling

The term "model" is often used to refer to some person or object—a person who shows-off clothing designs, or a small-scale replica of any of a variety of things. How does a model car differ from a toy car, though? How does a clothing model differ from someone who is simply wearing the same clothes? How does a mathematical equation differ from a quantitative model? The thing that makes a model unique is that it is a representation of something else—it is not the model itself that is of interest but, rather, the thing of which the model is intended to represent. A model of a car is an exact copy of a real car, except on a much smaller scale and generally does not function except as a copy of a car the model-builder enjoys. Clothing models are not the stars of fashion shows, the clothing designers are, and the models function as a representation of what people will look like wearing the designs being shown. Quantitative models exist as a representation of some system, mechanism, process, organization, or subject; they include exact measurements associated with the thing it represents, providing useful information about it to its user. A quality model will allow the user to not only derive information about a single moment of time in the existence of the thing being modeled, but will allow the user to run simulations under varying conditions, perform stress-tests to determine the limits of what is possible, and predict results. When observing a model of an organization, for example, each of the functions of that organization will be included, quantifying various aspects of its resources, operations, and networks, as if the organization was fully functional on paper, so that the users of that model can study the organization in great detail and determine the best way to manage their encounters with it before making any decisions. Such a model would function in a similar way to that of a map, offering guidance in navigating the operations, systems, and interpersonal networks within the organization. A quantitative model might show cause-and-effect responses to interventions, or predict the outcome of a particular course of event. The thing that they each have in common, though, is that they are each mathematical representation of things or events of interest.

As noted by Peter McCullagh, of the University of Chicago, on the field of applied statistics, "sound practical advice is understandably considered to be more important than precise mathematical definitions" (2002, p. 1225). The priority of a model is to be functionally useful, which often means sacrificing mathematical perfection. Each model is a ratio of effect (i.e., variance explained) and error (i.e., variance unexplained), wherein the existing model will calculate the size of an effect, and also include a variable representing the amount of variability in values that are not accounted for in the calculation. In the calculation for standard deviation, one of the steps is to subtract the mean from each value, square each difference, then add them together, which is known as the "sum of squared errors" as each observation deviates from the prediction of the model calculating the average; Σ(Observed – Model).[2] All the things that influence the thing being studied may never be 100 percent recognized, which means that 100 percent of the variability may never be fully understood. Through statistical sampling and measurements, there may be variability in the descriptions, creating a degree of statistical uncertainty as to the exact values the population being studied maintains. It is even in the nature of nearly all fields and aspects of our lives, both professional and personal, that we regularly make decisions without having all the information we would need for that decision to be fully informed. We use satisficing behavior, which means we accept when something is "good enough," and for models this means that it functions properly in a manner that is accurate enough for the user to make effective decisions, but without becoming paralyzed with indecision, uncertainty, and the endless pursuit of a goal in precision that may not be possible, at all. As a result, models generally include elements of statistics and probability, and elements that treat data which are missing or otherwise violate certain assumptions (discussed more in chapter 4). Models comprise a set of these statistical units (including units that incorporate the amount of error in the model and the statistical impact of that error), the covariate space in which these statistical units overlap in their influence, and a response scale that sets the amount of response the final output has to each statistical unit. This is then all placed within a context of functional purpose, incorporating the specific elements of the application being modeled. Analysts incorporate any of a variety of statistical tools and tests into otherwise known variables, constants, or relationships, which provide parameters for likelihood, performance, correlation, difference, or other variable information. In developing quantitative models, these statistical tools and tests make usable mechanisms of analysis, prediction, or operations, which would otherwise prevent effective decision-making due to their variable and uncertain nature that would make some components unknown if not for our ability to assess statistical values and ranges. The applied usage of statistical analysis to a variety of topics makes precise definitions nearly impossible. Rather, it is important to understand the mechanisms of statistical analysis to incorporate them into models as needed, and applying to them the relevant variables in a way that is useful for the circumstances.

Models are built from research; they use observations and measurements to form abstractions, which can be applied to a wide variety of similar circumstances meeting defined criteria. Models, research, and theory are all intertwined, each contributing to the other and, in turn, shaped by the others, as well, with research functioning as an intermediary between abstract theory and empirical, applied modeling. Research discovers information, which is incorporated into the overall knowledge set known as theory and, in turn, existing knowledge drives new research as researchers seek to expand the set of total knowledge, fill in missing gaps in our understanding, or challenge existing knowledge. The new information that is formed by research is applied to a variety of relevant fields to improve operations, incorporating them into representations of operational functions known as models and, in turn, the data that are collected on operations based on those models are analyzes and the finding incorporated into the research. As a result, models are born indirectly from theory and even other models, but primarily from research, and though research can be quite difficult to perfect, it is simple to understand. Research is nothing more than the careful analysis of measurements taken of those things around us. Ideally, these measurements would be performed on the entire population being studied, but often that is not possible so, instead, samples are usually taken. A population refers to the entirety of whatever is being studied, whether that is every single person on the planet, all the vehicles in a single side of a conflict, every occurrence of a particular type of event in the 20th century, or anything else; the population of each of these things includes every single member of that group. When measuring every single item in an entire population, the process is called a census, but these are very time-consuming, very expensive, and the members of the population are often unwilling or unable to participate. More common, then, is the use of a sample, which takes measurements of a portion of the population, then performs analytics known as inferential statistics to determine whether the descriptive statistics taken from the same match those of the population from which the sample was taken.

There are two broad methods for sampling: nonprobability and probability. For example, nonprobability sampling involves intentionally choosing specific participants from the population to measure. For example, it is common to study the people of the United States by using Omaha, Nebraska as a sample, because Omaha's demographics match very closely with those of the US population, and choosing Omaha is a deliberate choice of sample that involves no probability. Quota sampling is common, in which subjects being measured are chosen to fulfill a quote of particular traits, while judgment sampling uses the judgment of someone experienced in the field, but these are often subject to systematic bias errors and must be utilized only when there is some very specific reason for them. Generally, the more preferred method of sampling is probability sampling, which utilizes methods of choosing subjects from the population based on random selection. Simple random sampling is the most commonly cited method, and relies on any method that selects subjects completely at random from the entire population. Consider if you had each unit of the population written on individual slips of paper, which were placed in a hat, and you randomly pulled slips out—this would be a type of random sampling. Another way would be to randomize a list of the population and choose every nth one (e.g., every third in the list, or fifth of hundredth, etc.). If circumstances require that the sample include a minimum number of subjects of different specified groups, or which had specified traits, then utilizing stratified random sampling is generally the best method. This method divides the entire population into specific subgroups, and then randomly samples from each of those subgroups, either equally, or in weighted proportions that represent their proportion of the population or their proportion representation in the model. For very large populations, often multistage sampling is used, wherein increasingly narrow categorizations are sampled. For example, a specific geographic region might be sampled from all regions, then a specific group of people in that region might be sampled from the entire population in that region, then individuals within that group might be sampled from the entire group. These geographic samples are also categorized as a type of cluster sampling, which refers to the sampling of subjects within larger clusters, to avoid having to sample from an entire dispersed population, which can be extremely time-consuming and expensive. In the end, to get probability or nonprobability sampling right is a lot of work, so it is the unfortunate truth that far too often researchers rely heavily on something called convenience sampling, which refers to the process of sampling whatever subjects are conveniently available. Under dynamic environments, it is often difficult to get a proper sample or to measure the sample properly, so studies frequently utilize an element of convenience sampling since subjects must either participate, or at least not be aware they are being studied, but researchers try to minimize it as much as possible. This is increasingly becoming less of a problem, however, as technologies such as satellites, drones, computer and electronic reconnaissance, and other methods have made data much more available, and its collection much simpler, than at any other time in history.

The methods of data collection discussed so far have been merely survey methods, in which the population or a sample is merely observed to passively collect data. Another method of data collection that is far less common but much more useful, is experimentation. Experimentation, at its best, isolates subjects from all extraneous influences and exposes them to a treatment of some sort in order to measure their response. A military might conduct exercises, tests, or other shows of capabilities in order to gauge the response of the opposition, or they might move their own resources to a pre-planned destination in an attempt to create a response from the opposition that will provide useful information. In human research, it is common to measure the response of people to specific stimuli, sometimes while scanning them to measure their physiological response using fMRIs, PET scans, or other such equipment. A very big consideration with experimentation, though, is ethical. Specifically, how does one protect the safety and dignity of the participants? Even when performing experiments using the opposition, perhaps especially when experimenting with the opposition, it is important to maintain the highest degree of ethics and not do anything you would not do to voluntary participants for whom you have great appreciation for their participation. The backlash of unethical experimental research is severe, and includes such infamous people as Joseph Mengele, and events such as the covert radiation experiments performed on unwitting US citizens by their own government between 1944 and 1974. The methods with which data are collected can be a turning point in any conflict should allies and the public turn against you for using unethical methodologies.

When taking a sample, the size of the sample matters tremendously. Studying the entire population is preferred, if possible, but, if not, taking as large a sample as is prudent is necessary. Smaller sample sizes result is greater degrees of random error, as calculated by $E = Z(S/\sqrt{n})$, wherein Z is the z-score, S is the sample standard deviation, and n is the sample size. This calculation of random error is also utilized to measure something called a confidence interval, which calculates the reliability of an estimate by determining what range of values are included within a given confidence based on predetermined probability, calculated as x plus or minus the standard error. The law of large numbers states that an increasingly large sample will be increasingly representative of the population. Intuitively, this seems pretty obvious; of a population of 100 subjects, to take a sample of 1 subject would likely not represent the rest of the population very well, but a sample of 99 would include almost the entire population, and so it would be very representative. Take note of the importance of random sampling, however, in that taking a sample of 50 from the population of 100 will still be a bad sample if they are all intentionally taken from one side of the population, rather than evenly distributed. In order to eliminate the impact of sampling bias, random selection is used. Another implication of the law of large numbers is that the more samples taken will be more greatly representative of the population, on an analysis of those samples. In theory, sample sizes are chosen with a set statistical strength, variance, or error in mind, and the sample must be large enough to meet minimum requirements of strength or certainty. In practice, deadlines and funding tend to be just as important, but with a sufficiently large sample size, another interesting thing happens, described by something called the central limit theorem. The central limit theorem states that an increasingly large sample size will greatly resemble a standard deviation, when meeting certain parameters. So, a sufficiently large sample will have a mean equal to the population mean as a result of the law of large numbers, and the distribution of the data around the mean will be a normal distribution as a result of the central limit theorem. Suffice it to say, that getting as large a sample as possible is of the utmost importance, because these things will be strong determinants of the accuracy of any analysis, and the strength of statistical tests used.

There are a variety of different statistical tests that will be discussed during all three parts of this book, but what is important now is to understand how each can be incorporated into a model. As noted near the beginning of this chapter, models are composed of several pieces: the statistical units, the covariance between those statistical units, and the degree of response to which the calculation responds to each statistical unit. So, in seeking to calculate x, a very simple model might look like this: $x = \text{Cov}[a, b]/[a(y) + b(z)]$, wherein a and b are statistical units—either known values or the value generated by some other sort of calculation—while y and z are multipliers that determine the amount of weight given to either a or b in the calculation. $\text{Cov}[a, b]$ refers to the degree to which a and b vary in their values together, rather than idiosyncratically, which is called covariance. Just as a and b might be any known value or measurement, statistical tests can also be inserted there, depending on what you are trying to accomplish. Granted, it is not the test, itself, that is important, but the value calculated using that test, but that is not the nature of a model. A simple volatility model looks like this: $\text{Cov}[s, g]/\sigma_g$. Standard deviation is a calculation on its own, but since the standard deviation is not yet known, it has been replaced with the generic representation of its value. The same goes for statistical tests—a model might require that the amount of correlation between two variables be tested, but until that value is calculated, all that can be done is to replace it with a representation of that value, which is the generic model for the test, itself. To simplify things, this is usually done by replacing the entire calculation of the test with a single variable, as was done with standard deviation in the simple risk model from earlier. There is no single way to arrange the elements of a model, as it is all entirely dependent on the thing being modeled. The important thing is to capture the functionality of the thing being modeled in a quantitative form, which can be used to properly analyze that thing.

Quantitative models are generally represented in two ways: equations and graphs. Equations are used to make calculations, and to determine the precise value of whatever is being determined; it is the more functional and accurate of the two forms a model can take. Many people have difficulty visualizing or even understanding equations, however, which is why graphs and charts are frequently used to visually illustrate the relationships of the values being included in the equation. The type of graphical representation used depends entirely on the equation and the type of information being communicated. One of the most common types of models describing the distribution of data, graphically, has already been described in chapter 1 on

"Descriptive Statistics": the bell curve of a normal distribution. This curve is actually derived from something called a histogram, which graphically represents the frequency with which a specific value is encountered in a data set. Using vertical lines, the length of each line represents the number of times that value was encountered, and longer lines indicate more common values, as described in chapter 1. The curve, itself, represents the distribution of the data, visually providing information about the data set, including the average, skew, kurtosis, and so forth. As with all models, this information can also be represented as a mathematical equation, offering information that is more precise and useful, but which can be more difficult to visualize and process. The equation for a normal distribution, for example, is as follows:

$$f(x, \mu, \sigma) = \frac{1}{\sigma\sqrt{2\pi}} e^{-\frac{(x-\mu)^2}{2\sigma^2}}$$

Note that the normal distribution is a function of its mean and standard deviation, just like in the graphic version, but that equation is simultaneously more detailed yet more complicated. This trade-off between equations and graphs is the reason why it is so common to produce both when building a model, so that the best traits of each can be made available—the equation for when calculations are being made, and the graphs for when an analyst needs to communicate the calculations to others.

There are many different types of models. Range and IQ range are best illustrated using box plots, which are very useful is graphically illustrating statistical tests comparing differences using t-tests, some ANOVAs, and other types of comparative tests discussed in this book, particularly in chapter 3 on "Comparative Assessments." These range-based models can also be used as an element in time-series models, if they are used to measure the same thing multiple times and compare changes over time. This is extremely common in tracking both a moving average, as well as tracking upper and lower values of resistance, which are the maximum and minimum values that volatile data tends to reach before slowing the rate at which it changes over time, and frequently changes directions. These are often illustrated using a graphical model known as Bollinger Bands, which is a time-series model comparing the moving average with a range of data that is plus or minus any predetermined number of standard deviations from the mean, offering a relative definition of "high" and "low" in ranged, time-series models.

Changes in some value in response to another can be modeled using trend-line computations, which refers to any calculation that measures trends in the value of any variable. That may include changes in the value of some variable over time, or in response to the influence of another. For example, the strength of a battalion may increase over time, with experience and training, and it may also increase as a function of the volume of specific types of resources that are available to them. Trends such as this are simplest when calculated linearly, that is to say, as a straight line which, as you may recall from introductory algebra, is calculated using the equation $y = mx + b$, wherein y is the variable value being forecasted, b is the value of the dependent variable being modeled when the independent variable being used as a predictor is 0 (also known as the y-intercept), and m is the slope of the line. The slope of the line is a calculation of whether the value is trending up or down, and how drastically it is changing; a steeper line indicates more quickly changing values. The slope is really very simple to calculate; $m = (y_2 - y_1)/(x_2 - x_1)$. All this means is that the location and slope of the line that graphs the relationship between variables (including changes in a dependent variable over time, in which case time is an independent predictor variable) can be measured using 2 points on the graph. Then, by comparing the difference in the y and x values of those two points, the slope of the line can be calculated.

Of course, in statistics it is often the case that a line modeling the shape of an average is being calculated while the values, themselves, actually deviate from the average, eliminating precise reference points using that simple slope model. In this case, a statistical slope model must be used, such as the simple least-squares approach, which is more time-consuming but not any more difficult: $m = [(\Sigma xy) - (\{\Sigma x \Sigma y\}/n)]/[\Sigma x^2 - (\Sigma x)^2/n]$. There is a reason why these things are all calculated by computer, now, and it is not that it is very difficult, but that it is very time-consuming when using large data sets, requiring a person to multiply or sum huge volumes of values. Still, this only functions with linear models.

Many models are known as curvilinear, which means that they can be represented by a single line, but one which curves rather than being a straight line, and there are several different types. Logarithmic curves ($\log(x)$) increase in value but at a slowing rate, while exponential curves (x^y) increase in value at an increasing rate. Partial derivatives, which will be discussed in more detail in chapter 13 in regard to the marginal rate of substitution, hold one independent variable constant, which creates a variable influence on the remainder of the calculation resulting in predictable patterns in the shape of the curve, which are very common among economic analytics associated with the rates at which resources are produced or consumed, the value generated in investing resources toward particular projects, and so forth. An extremely basic type of curvilinear model commonly utilized in a variety of fields is known as an elasticity model, which has a core equation model of $\Delta x/\Delta y$, wherein x and y are whatever variables being studied, and Δ is the symbol delta, indicating rate of change as a percentage. These measure specifically how the value of one variable will change in response to a change in another variable. They are very simple to calculate and can be used to measure the proportion of response, which just about anything has to anything else. Just start with the percentage change of the independent variable, then measure the percentage change in the dependent variable, which happens as a result. For example, you could start with a percent increase in the total volume of people and resources being mobilized, and measure the amount of time required to complete the mobilization process in response; you could measure the percentage response in opposition morale to change in the percentage difference of troops, which could be modeled as: $\Delta d_t/\Delta m_o$, wherein d_t is the amount of differential of troops in an imminent combat situation, and m_o is the amount of morale maintained by the opposition. These measures of elasticity tend to be curvilinear, as shown in the scatterplot in figure 2.1.

Wherein the rate of response, itself, changes. In this case, as the differential of troops increases, opposition morale drops, but it drops

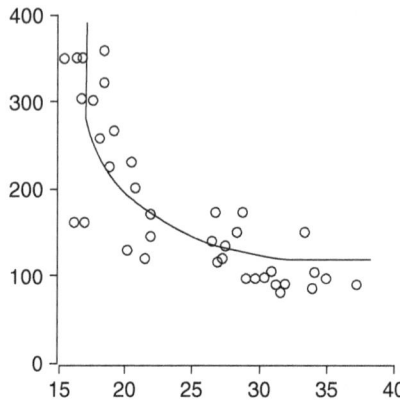

Figure 2.1 Curvilinear scatterplot

at a decreasing rate, as a result of something called the law of diminishing marginal utility. No matter how big the differential in troops gets, there will be a point at which the entire opposition simply knows they cannot win, leaving those that will stay to fight for the glory or martyrdom of dying in battle while defending their side of the conflict, and so morale will stabilize at a given point. Each two points on that curve has a given slope, but the slope of the curve changes for each two points; the slope on the left side of the curve is much steeper than that on the right. The rate at which the line changes its slope is called the marginal rate of transformation. The exact equation for these graphical models depends on the type of curve, whether it is exponential, logarithmic, square root, inverse, sin/cos/tan, or any of a huge variety of other options.

One type of curve that is particularly useful when modeled is called a parabola, which illustrates an upward or downward trend that reaches some maximum or minimum point, respectively, and then reverses direction, forming the shape of a U, perhaps upside-down. When collecting data, should the plotted data form this shape, the reasons it is useful is that it allows for a clear calculation of some maximum, minimum, or optimization point for which one can strive in achieving the pinnacle of whatever it is you are trying to accomplish. A very simple parabola is calculated as $f(x) = ax^2 + bx + c$, wherein the lowest point, called the vertex, is calculated using the equation $-b/(2a)$. This allows for a precise identification of the minimum possible value of, for example, the lowest average resource allocation required to successfully maintain operations in a particular mission, so that any surplus resources can be better utilized to create value in other areas or other operations, increasing the total effectiveness and efficiency of the available resources. One particularly common type of parabolic curve that is seen in social models quite frequently is the j-curve, which is so-named because it drops for a time before rising back up to levels that surpass the original level. In the reverse, the inverse j-curve does just the opposite, rising temporarily before dropping below the original point. The j-curve holds particular importance in models that illustrate national stability as a function of political openness, as explained by Ian Bremmer (2006), as well as in modeling domestic civil unrest, as explained by James Davies (1971).

Another type of model that will be discussed in more detail in chapter 7 is called probability modeling. These types of models function by calculating the likelihood of some observation being made, such as the probabilities associated with dropping cluster munitions, wherein a single bomb will split-up and drop smaller explosives. In a scenario wherein a cluster bomb is dropped over a target with a given population density x, and 20 mini bombs are dispersed over the area, what is the probability that an opposition soldier will be struck? In these sort of examples, the probability was pretty low, and the result ended up being that large numbers of undetonated small explosives would be lodged into the ground, causing a severe hazard for decades, which is why cluster munitions have been largely banned around the world.

Entire systems can be calculated using models, with each element given its own designation within the model using a sort of polynomial arrangement; x = (model of element 1)(model of element 2)(model of element n). For example, consider calculating the functional productivity of any organization; by breaking down the organization into its various functions, each function can be modeled, and then the models of each of those functions can be inserted into much larger organizational model. These can include known values and calculations, or statistical tests, as needed. Even if a particular value is not known, and there is little basis for calculating a value statistically, probability modeling can be used to evaluate the likelihood of particular values, so that almost anything can be modeled to a great degree of certainty, within a range of values. Models are built to be representations, and they can be custom-built to represent just about anything so that every aspect of a combat environment can be studied in great detail before making any decisions—allowing for the calculation of likely outcomes, possible outcomes, and tests for what limitations exist—allowing those who have functional models to develop the highest quality strategies and know the outcome of those strategies well in advance.

CHAPTER 3

Comparative Assessments

As stated by Sun Tzu, "The ground gives rise to measurement, measurements give rise to assessments, assessments give rise to calculations, calculations give rise to comparisons, and comparisons give rise to victories." Comparative assessments are a specific categorization of tests, which are used to determine whether two or more data sets are significantly different from each other. For example, in measuring the operational performance of two similar groups, the data may indicate that the two groups have different average performance, different performance ranges, deviations, skew, and so forth, but by just utilizing these simple descriptive statistics described in chapter 1, the possibility exists that the differences between them are not statistically significant. There may be differences in the samples that were pulled from each group, but those differences may not be present in the populations from which those samples were pulled. One group may have a mean that is slightly higher or lower than the other but with a much larger range, or the standard error of each group may be greater than the differential between the two means, making it impossible to determine whether there is actually a difference in performance between the two organizations through casual observation. Even if it appears obvious that a difference exists, taking into consideration all these errors and overlaps associated with sampling and comparison, the amount of difference that exists can be very difficult to determine. For example, looking at the enemy lines along the battlefront, it may seem that some geographic areas are thinner—less densely occupied with defensive resources—but whether the difference is great enough to make a significant impact on their combat strength, or whether that difference is great enough to be incorporated into a strategy that targets the weak spot, is something that cannot be determined with any certainty through simple observation.

During those times we must be certain that the differences between two or more things being compared are truly indicative of something significant, or when the amount of significant differences must be assessed, then there is a variety of specific types of tests that can be performed. The type of test performed depends on those things being compared; the types of data being analyzed, the number of different things being compared, and whether the data are parametric or not. In this chapter we will explore the various analytics that are used to test for differences in data, including those things which determine the tests that need to be used under different circumstances. There is wide range of analytics that can be performed to compare the differences among things with extremely high precision, and also a number of methods used to compare changes in a single thing over time, the results of both are utilized into predictive and operational analytics, and incorporated into models for each. Until one understands one's self and one's opponent, they are blindly charging into conflict. The data necessary to understand the exact nature of each player are available, and they can be used to accurately develop the optimal strategies to pursue, otherwise each strategic decision is being made arbitrarily.

Like most good things in this world, the most basic of comparative assessments has its origins in beer. During the first decade of the twentieth century, Guinness had in its employment a talented chemist named William Gosset, who developed an analytical method for comparing the quality of various batches of beer. To protect proprietary information, Guinness policy stated that chemists could not publish their research, so he used the pseudonym "Student," resulting in the test being forever named "Student's T-Test," which is colloquially shortened and is most commonly referred to as a t-test. The t-test is used to compare two groups of continuous variables, when the data are parametric. In the case of William Gosset, he invented the test so that he could take samples of different batches of beer, take quantitative measurements of several specific traits (e.g., alcohol content, viscosity, etc.), and then use the t-test to determine whether any differences in the measurements between the samples was the result of something significantly different in the batches, or simply minor fluctuations associated with sampling, or otherwise too small to have an impact.

Like all comparative assessments, there are two types of t-tests. The first, is the independent means t-test, which was the one developed by Gosset. As the name implies, the value of the two things being compared are independent of each other. For example, the combat capabilities of both sides in a conflict can be measured to determine whether there is a significant difference in strength between them, but each is not determined by the other. The independent means *t*-test is calculated as follows:

$$t = \frac{\overline{x}_1 - \overline{x}_2}{\sqrt{\left[\frac{(n_1-1)s_1^2 + (n_2-1)s_2^2}{n_1 + n_2 - 2}\right]\left[\frac{n_1 + n_2}{n_1 n_2}\right]}}$$

The value of the test can be positive or negative, but since it is only difference for which the calculation is testing, and that is inherently nondirectional, whether the test is positive or negative does not matter. If the test value exceeds the critical value, it is considered statistically significant, and we reject the null hypothesis. In other words, if the absolute value of the test value is greater than the absolute value of the critical value, then the probability that the difference is the result of random chance is too small, so we reject that idea. The critical value is the benchmark value to which we are comparing the test value. The critical value of a *t*-test is determined by the *p*-value and the degrees of freedom. The *p*-value is the probability that the researcher chooses as the level of significance at which random chance is no longer acceptable as a possible explanation. Common *p*-values are 0.05, 0.01, 0.001, and 0.0000003, which means that the probability of the test being the result of chance is less than 5 percent, 1 percent, and so on. The degrees of freedom (*df*) refer to the number of values in the final calculation that are free to vary; in the case of the independent samples *t*-test, the df is calculated as $n1 + n2 - 2$. This means that the sample size, n, minus the number of samples is the df. If there is only one sample, then $n - 1$ is the *df*, but since the *t*-test requires two samples, then adding the sample sizes together and subtracting 2 is the proper application of *df*. Once you know the *p*-value and the *df*, then you can determine what the critical value is by referencing the table in appendix B of this book. The nonparametric alternative to the independent samples *t*-test is called the Mann-Whitney U test, calculated as follows:

$$U = N1 \times N2 + NX \times \left(\frac{NX+1}{2}\right) - TX$$

The other type of *t*-test is called a dependent samples *t*-test. The difference is that the independent samples *t*-test compares two samples that are independent of each other, while the dependent samples *t*-test compares the values of the exact same sample before and after some intervention. In other words, by testing the combat capabilities of a battalion before and after some new form of training, it is possible to determine whether any changes were the result of the new training, or whether it was the result of chance. This helps to resolve questions about the possibility of whether soldiers were simply having a good/bad day at some point during the assessment, for example. The dependent samples *t*-test is calculated as follows:

$$t = \frac{\sum D}{\sqrt{\frac{n \sum D^2 - \left(\sum D\right)^2}{n-1}}}$$

In the case of dependent samples *t*-test, the *D* refers to the difference in score pairings. That means, for each individual person, you calculate the difference in their score from before and after the intervention, and then for each case you add them up; sum the differences, sum the squared differences, or square the sum of the differences. Notice that the *df* in this case is calculated as $n - 1$, since there is only one sample but it is being measured multiple times. The nonparametric alternative to the dependent samples t-test is called the Wilcoxon Signed Rank test.

In situations that are similar to those necessitating a *t*-test, but in which there are more than two samples, a *t*-test will not be sufficient. In these cases, an analysis of variance (known as ANOVA) must be performed. The ANOVA utilizes the calculation of the *F*-value, which is the ratio of the mean squares of the model divided by the residual mean squares, or MS_M/MS_R; in other words, the amount of variance explained by the model divided by the amount of variance left unexplained. These are calculated by dividing the respective sum of squares by each appropriate df. MS_m, for example, uses the between-group estimate as its *df*, giving it a calculation of $SS_M/(k-1)$, while MS_R uses the within-group estimate, giving it a calculation of $SS_R/(N-k)$, wherein *N* is the total number of samples. The respective sum of squares is calculated as follows:

$$SS_M = \left(\sum\left(\sum x\right)^2 / n\right) - \left(\left(\sum\sum x\right)^2 / n\right)$$

$$SS_R = \sum\sum(x)^2 - \sum\left(\sum x\right)^2 / n)$$

When two sigma symbols are placed next to each other consecutively, it means to sum all the values between groups, so $\Sigma(\Sigma x)^2$ means to sum of *x* in each sample, then to add them together. Therefore, to summarize basic ANOVA, start by calculating the sums of squares, then use those to calculate the mean squares, and use the mean squares to calculate the *F* value. With the *F* value, use the chart in Appendix C to determine whether the *F* value exceeds the critical value.

When used to determine whether there is a difference before and after treatments, as with a dependent means t-test, the same thing is done but using a single sample, calling it a repeated measures ANOVA. A repeated measures ANOVA is useful for tracking changes over the course of a regular treatment, such as the implementation of an annual training program, should one want to measure whether the new training changes in efficacy over the course of several years, rather than just two years. The nonparametric alternative of the ANOVA is called the Kruskal-Wallis test, as shown:

$$\frac{12}{N(N-1)} \sum_{i=1}^{k} \frac{R_i^2}{n_1} - 3(N+1)$$

There are several variations on the ANOVA, each used within a specific context. When there are multiple independent variables, rather than just the one categorical independent variable as would be appropriate for an ANOVA, then a factorial ANOVA, or FANOVA, is performed. The FANOVA is calculated and functions exactly the same way as a standard ANOVA, but ensuring to account for the increased number of independent variables. When a standard ANOVA has an extra independent variable associated with it that functions as a predictor of the test outcome, and varies in relation to the primary independent variable, an Analysis of Covariance (ANCOVA) is most appropriate, because it allows a person to control for independent variables in order to determine how much idiosyncratic influence each has. This is mechanically similar to a multiple regression (discussed further in chapter 8). ANCOVA is calculated a little differently, to neutralize the impact of the covariate; linearly:

$$Y = GM_y + \tau \left[B_i \left(C_i - M_{ij} \right) + \cdots \right] + \varepsilon$$

Factorial and covariate variations also exist for the final type of ANOVA that will be discussed in this chapter: MANOVA. The multivariate analysis of variances, or MANOVA, is used in the same context as a standard ANOVA, except that there are multiple dependent variables. The factorial MANOVA also has two or more independent variables, and incorporates into a MANOVA the same elements that FANOVA does for ANOVA. Should a MANOVA have a covariate, then just like an ANCOVA, the MANOVA can be turned into an ANCOVA by controlling for the covariate. Just as FANOVA and ANCOVA have a standard ANOVA at their core, so too do both the factorial MANOVA and the MANCOVA have the same core analysis: the MANOVA. That being said, as already noted, MANOVA is a comparative assessment with two or more dependent and independent variables. This helps to identify interactions between the independent variables and dependent variables, the interactions between independent variables, and the interactions between dependent variables. For example, assessing the comparative combat capabilities of multiple groups in a combat theater, each composed of different types of equipment, soldiers with different types of training, and different measures of capability. A MANOVA is calculated using the same basic mechanism as an ANOVA, wherein the within and between groups sum of squares, and calculates significance as a function of the model variance and residual variance, but rather than using a standard F value, more commonly used is Wilks' Lambda, Λ, calculated as follows:

$$\prod_{k=1}^{m} \frac{1}{(1+\lambda_k)}$$

The tall symbol on the left is the capital letter Pi, and means multiply each value of the following specification. It functions like Σ, except that it says to multiply rather than add. The A MANCOVA, then, would measure the same, but while controlling for covariates.

The final test of comparative assessment this chapter will describe is special in that it does not include any continuous variables—only categorical ones. This test is called a Pearson Chi-Squared test, χ^2. The chi-squared test is calculated as follows:

$$\sum \frac{(O-E)^2}{E}$$

Wherein O is the observed frequency, and E is the expected frequency. For the expected value, the total number of participants can be divided equally among all potential responses (with 100 participants and 4 answer choices, an equal expected distribution would be 25 each), or they can be weighted in whatever manner is appropriate for the analysis. Since all variables are categorical, only the frequency of observation is measured. For example, taking samples of the types of human resources in each side of a conflict to determine whether there

Table 3.1 Reference guide

Independent Variables	Dependent Variables	Covariates	Type of Analysis
Categorical	Categorical	0	Chi-Squared
1 Dichotomous	1 Continuous	0	t-test
1 Categorical	1 Continuous	0	ANOVA
1 Categorical	1 Continuous	1+	ANCOVA
2+ Categorical	1 Continuous	0	Factorial ANOVA
2+ Categorical	1 Continuous	1+	Factorial ANCOVA
1 Categorical	2+ Continuous	0	MANOVA
1 Categorical	2+ Continuous	1+	MANCOVA
2+ Categorical	2+ Continuous	0	Factorial MANOVA
2+ Categorical	2+ Continuous	1+	Factorial MANCOVA

are significant differences in the types of skill sets being represented so that strategies can be customized based on those differences would qualify as an appropriate context for the use of a chi-squared test.

See table 3.1, which is a reference guide that will help to identify when to use each type of comparative assessment.

Besides the p-value and test statistic, another value that can be calculated that provides insight into the comparisons of those things being measured is the effect size. There are several different types, each appropriate under different circumstances, but each utilized to accomplish the same thing: evaluate the strength of the association. The size of an effect ranges between 0 and 1; values around 0.2 indicating a small effect, 0.5 a moderate effect, and 0.8 a large effect. In the case of comparative assessments, there are three different effect size calculations, which are commonly used, each used to determine the amount of overlap of data distribution between samples in order to calculate the degree of difference between them. This differs from the comparative assessments, themselves, in that the comparative assessments are used to determine whether the difference is significant, while the effect size is an estimate of how much difference there is in the data without actually calculating whether the difference is enough to warrant the rejection of the null hypothesis. Test statistics and effect sizes are used in conjunction to provide a comprehensive assessment of the comparative traits of two or more groups.

When estimating the effect size for comparative samples, the core equation takes the difference of the means between the two groups, then divides it by some measure of deviation. Glass's Δ, for example, is calculated as $(\mu_T - \mu_C)/\sigma_C$, and is used when comparing a test group, T, against a control group, C. An effect size calculation that is more versatile, and as a result more commonly used, is Cohen's d. What makes it more versatile is that it is bidirectional; rather than measuring against a control group, it measures the difference between two groups using a pooled standard deviation, as follows:

$$\frac{M_2 - M_1}{\sqrt{\frac{(n_1 - 1)s_1^2 + (n_1 - 1)s_2^2}{n_1 + n_2 - 2}}}$$

Note that the numerator is still the difference between the means of the two groups, but that the denominator now incorporates the root of the standard deviation of both groups as a ratio of the degrees of freedom. The effect size calculation for ANOVAs is called the Root Mean Square Standardized Effect, RMSSE, or Ψ:

$$\sqrt{\frac{\sum_{i=1}^{n} \frac{\alpha_k^2}{\sigma}}{k - 1}}$$

This calculation allows for k volume of groups being compared, which can be as many as needed.

Comparative assessments are critical when assessing the relative strengths and weaknesses of the players in a conflict. They can be used to determine the outcome of each encounter, though they cannot quantitatively define the amount of resource usage that will be associated with those encounters. To accomplish that, another type of assessment needs to be done, called regression, which is discussed in greater detail in chapter 8. Still, comparative assessments can be used to test for variations in geographic distribution to identify points of particular strength or weakness. They can also be used to assess changes within a single organization over time; when incorporated vigilantly, it can be used to identify specific practices and behaviors within the organization that lead to successful outcomes, improved performance, and increased competitiveness. It is very difficult for the casual observer to determine whether what they have observed is indicative of a significant difference or whether subjective error is inherent in their observation, much less whether their observation is representative of the entire population, most of which will likely be unknown to them. In order to determine with certainty whether two things are comparatively different in a manner that is great enough to cause a significant outcome, they must be assessed against each other by assessing measurements made from each of them.

CHAPTER 4

Data Diagnostics

The majority of statistical tests and analytics are performed under a variety of assumptions about the descriptive distribution of data, the influences that determine the values of the data, and the manner in which the data are collected. Assumptions play an important role not only in analytics, but in daily life; a person generally makes the assumption that what he sees is representative of reality, or that orders given are being properly followed, or even that six nuclear warheads could not be accounted for one moment and then simply disappear in the next. These assumptions are based on patterns necessary for even simple functioning—if we did not assume that the ground beneath us would not fall away as we walked, then each person would be paralyzed with uncertainty. During planning, it is assumed that any orders given will be executed correctly and in a timely manner, otherwise the desired goals will not be accomplished, and the unreliability of any particular element within the organization makes management of that organization nearly impossible. When commanding a unit to protect the nuclear warheads at Minot Air Force Base in North Dakota, if command does not assume that this will be performed properly, then they cannot build upon that action or take advantage of the goals they intended to accomplish. Perhaps, war is declared and issues are ordered to launch these nuclear warheads at opposition forces, but if the warheads are not available because of a violation in the assumption of their proper storage and handling, then the planned launch cannot be executed. In the same way, there are certain assumptions that we rely on in our statistical analytics, which generally tend to hold true, particularly with very large data sets. Certain statistical tests require the data to exhibit particular traits, or else another form of test must be relied upon. Without thinking, we often make inferences that tend to be less than accurate—we look at averages far too often as if the mean is truly centered without checking, while the data may actually include a rather severe skew. When we use assumptions, they must be true in order for the analyses we perform to yield useful information, but that does not mean that this is done recklessly—there are tests that can be performed to determine whether the data being analyzed violate these assumptions and, if they do, there are methods of manipulating the data so that it can be made useful. Collectively, the process of testing assumptions and manipulating data are known as data diagnostics, because in doing so you are diagnosing the data for violations of assumptions.

Many people use the word "assumption" interchangeably with "guess," but the two are very different things. An assumption is a parameter that must be maintained as a foundation for logical deductions to retain functionality, while a guess occurs when one simply accepts that assumption to be true without evidence. In planning, command will assume that orders will be executed correctly and continue to make plans based on that assumption before they are ever put into motion, but many steps are then taken to ensure that the assumptions used remain intact. Controls are implemented to ensure that orders are executed, such as drills and evaluations, and operations are monitored so that a proper response can be made to deviations in the assumptions that might arise. Just as when operational assumptions such as this are encountered management will respond to ensure that the plans built on those assumptions will still be executed, when a violation of the assumptions of data are violated there are several responses that can be taken to accomplish the goal of extracting useful information. At no point is an analyst merely guessing that these assumptions are true, but they test the data to ensure it is true and, if they are not, respond appropriately.

The most common and basic of statistical assumptions is normality, meaning that the data maintains the traits of a normal distribution, as described in chapter 1 on "Descriptive Statistics." It is extremely important to be aware of whether your data are normally distributed because not only do many statistical tests depend on this assumption being true, but even basic descriptive inferences such as mean, range, and standard deviation can be extremely distorted if this assumption is violated. There are several different ways in which the assumption of normality can be violated, one of the most commonly cited being the existence of outliers. An outlier is a value that exists far above or below the bulk of the data set. In other words, it is a single value that whose value is so extreme compared to the rest of the data that it appears as though it should be impossible. In fact, it is the definition of an outlier that it must have less than a 0.05 percent probability of being encountered, meaning that an outlier is any value that has a z-score of 3.29 or greater. The existence of an outlier will skew the distribution of the data, causing a shift in the value of the mean. For example, with a data set {5, 5, 5, 5}, the mean is obviously 5, but with data set {5, 5, 5, 25}, the mean is now 10 because of the existence of an unusually high value in the data set. If this data were related to the distribution of opposition resources in a combat theater and did not look at the raw data, reading only that the mean was 10, then it is likely that they would assume that the distribution was relatively homogenous across the region, while in reality there is one area that has an extremely high volume of resources. This would leave any strategic plans woefully incorrect, either leaving forces unprepared to defend against the area wherein the opposition has high concentration of resources, or leaving one completely unaware that other areas are potentially underdefended and able to be overtaken. Since both the mean and the standard deviation are distorted in data sets, which contain outliers, many statistical tests become invalid, so different tests known as non-parametric tests, for data which does not maintain statistical parameters, must be performed, but even using these tests that are more robust with the existence of violated assumptions than standard parametric tests, the results of the data will still be skewed and the test results are likely to be faulty. As such, the data itself must be treated.

For people not familiar with statistical analysis, altering data often sounds an awful lot like lying, making the data completely dysfunctional. Under most circumstances, they would be correct, but under certain conditions it is appropriate to use specific types of data alterations. It is appropriate when the alteration is the lesser of two evils. In others words, when the violated assumption is severe enough, it is creating more of a distortion to the data than the alteration. In the case of an outlier, there are several options, the simplest of which is simply removing the outlier. This can be done because the probability of observing the outlying value is practically 0 percent, besides the possibility of the value being the result of some mistake. The exact degree of impact of removing the outlier can be measured by assessing the change in slope of a cumulative probability curve, or regression slope (discussed more greatly in chapter 8), at which point the impact of the outlier can be precisely determined. Still, the complete removal of a value that is particularly high or low causes a higher impact than merely changing the value, such as taking the second highest value and adding 1 to it ($x+1$), using the value that is 2 standard deviations from the mean, or has a z-score of 3.28 (technically not an outlier). These changes create less of an impact on the data than the outlier itself, by decreasing the data distortion while maintaining a particularly extreme value that does not have the unfortunate trait of being virtually impossible to observe that outliers maintain.

Outliers are the simplest violations of normality for which an analyst can test, but data sets can deviate in ways that are less conspicuous, such as a skew or kurtosis that deviate too greatly from 0. Testing data sets for normality, and testing for the normality of a population from which a data set is derived, requires some more complex analysis, which compare the data to a reference point, normal distribution, and testing whether there is a significant difference between the two. Visually, this can be modeled using P–P plots or Q–Q plots, the latter of which is shown in figure 4.1.

Both types of plots compare the observed cumulative distribution of values to a reference point—in this case, the cumulative distribution

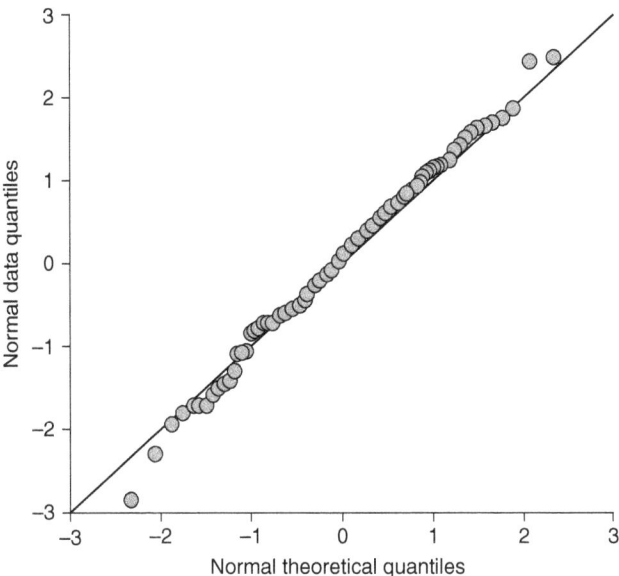

Figure 4.1 Q–Q plot

of a normal distribution. These visual representations can provide a simple way to casually determine whether there is some glaring deviation from a normal distribution, and function of a convenient method of double-checking for errors in more accurate, quantitative tests for normality. Such tests include the Shapiro-Wilk Test, which tends to be one of the most common:

$$\frac{\left[\sum_{i=1}^{n} a_i \left(x_{n+1-i} - x_i\right)\right]^2}{\sum_{i=1}^{n} \left(x_i - \bar{x}\right)^2}$$

Though a variation on the Kolmorogov-Smirnov test, which was discussed in chapter 3 on "Comparative Assessments," known as the Anderson-Darling Test, has been shown to be somewhat more accurate:

$$-n - \frac{1}{n}\sum_{i=1}^{n}(2i-1)\left(\ln \Phi\left(Y_i\right) + \ln\left(1 - \Phi\left(Y_{i+1-y}\right)\right)\right)$$

These tests are commonly pre-programmed into analytics software packages, and are rarely, if ever, actually calculated manually. These tests also require a certain degree of prerequisite knowledge of mathematics, which goes beyond what is expected of the reader of this book, so we will not be going into great detail into the mechanics of how these tests work, or how to perform them, but we will be introducing each just briefly to help introduce you to what they are, so that you might become better acquainted with them. They are both just ways of accomplishing something very simple: testing the distribution of data compared to a normal distribution to see if there is a significant difference, just like the comparative assessments described in chapter 3, except using the normal distribution as a reference point rather than testing for differences between samples. These are still limited to univariate analytics, however. To test for normality in multivariate statistics, a different test known as Mardia's Statistic must be used, which tests for skew and kurtosis for deviations from normality.

Another assumption that is frequently made is that of the homogeneity of distribution. It is natural behavior to presume symmetry and consistency in the density of the distribution of data. Think of data mining just like normal mining—it is easy to assume that when you start digging the ground will gradually get denser, at a relatively even rate, as you get deeper—just as the data in a normal distribution gets denser as you get nearer to the mean, as an even rate. Consider, though, that as you are digging that you find bit empty spaces, and other spaces that are super dense. Another type of homogeneity occurs when there is a consistent relationship between variables, say, between the amount of equipment being moved and the speed with which it can be transported. If the data are homoscedastic, it does not necessarily mean that there is a stronger relationship between these two variables, but that the strength of the relationship will remain consistent. When a relationship like this is heteroscedastic, a violation of statistical assumptions, it means the strength of this relationship changes over the sample; for example, as the volume of equipment increases, speed may drop but the amount of consistency in speed might also decrease, resulting in gradually greater variability. Tests such as Goldfeld-Quandt,

$$GQ = \frac{\hat{\sigma}_1^2}{\hat{\sigma}_2^2} \sim F(n_1 - k_1, n_2 - k_2)$$

and Breusch-Pagan (which is a process, not included in this book) check for violations of homoscedasticity, while Levene's test checks for generalized homogeneity.

$$W = \frac{(N-k)}{(k-1)} \frac{\sum_{i=1}^{k} N_i (\overline{Z}_{i\cdot} - \overline{Z}_{\cdot\cdot})^2}{\sum_{i=1}^{k} \sum_{j=1}^{N_i} (Z_{ij} - \overline{Z}_{i\cdot})^2}$$

Bartlett's test assesses homogeneity of variance among univariate calculations by evaluating whether there is equal variance among individual variables.

$$T = \frac{(N-k) \ln s_p^2 - \sum_{i=1}^{k}(N_i - 1) \ln s_i^2}{1 + (1/(3(k-1)))\left(\left(\sum_{i=1}^{k} 1/(N_i - 1)\right) - 1/(N-k)\right)}$$

Box's M tests is actually the same thing as Bartlett's test, and is even calculated in basically the same manner, except that it is utilized in multivariate calculations. All these things measure the assumptions within samples, but sometimes in repeated-measures designs that include taking several samples and comparing them, there will be heterogeneity between the various samples, known as a violation of the sphericity assumption, which can be tested for using Mauchly's Sphericity Test.

Depending on whether there are violations of normality through simple skews and kurtosis, or whether there is distribution variability, once they have been identified using the appropriate tests, there are a few options available that can help to resolve these assumptions of violations. One way is to simply use a nonparametric test of the data. The strongest tests are the ones that utilize these assumptions that have been discussed throughout this chapter, but there are often other tests available which are robust to violations of assumptions, but which are generally not as accurate or useful as their parametric siblings. It is also possible to apply transformations to the data set, which alters each value in the data set using an adjustment that alters values differently. For example, when a data set has a strong or increasing skew, it can be prudent to use a square root transformation, wherein the square root of each value is used ($\sqrt{x}$). Since square roots affect large numbers more greatly than small numbers, this square root transformation is used as a method of adjusting for extreme values. The opposite of a square root transformation is a squared transformation (x^2), or any other type of power transformation (x^y), used to spread out data. Another type of transformation that can be used is the log transformation ($\log(x)$). Since logarithms increase in value at a decreasing rate, taking the log of each value is another way to more greatly adjust extreme values, so that the more extreme a particular score is, the more it is adjusted back toward the mean, similar to a square root. A reciprocal transformation ($1/x$) brings values toward 0, reversing their order (100 becomes 1/100, while 10 becomes 1/10, which means that numbers which were smaller before the transformation are larger after the transformation). This reversal can be accounted for by utilizing a constant in the transformation, though, so that the reciprocal includes subtracting each value from the highest value, $1/(X_{highest} - X)$. Using a constant in the transformation is also necessary when working with negative numbers, so that all the values can be made nonnegative before the transformation, and then adjusting again, afterward. The number of types of data transformations that can be utilized are theoretically limitless—the important thing is to know exactly what the problem with the data is, and how you can adjust it. So long as the transformation is applied to every value in a dataset, the adjustment will not invalidate the analysis, as can simply removing values.

Removing values is a simpler method of working with problematic data sets, however, and not necessarily harmful to the results of an analysis. Trimming data refers to removing a small percentage of the data at either end of the data set, such as 5 percent or 10 percent of the total data, to work exclusively with the data that composes the vast majority of those values that are likely to be encountered. Recall from chapter 1 on "Descriptive Statistics" that interquartile range measured the range of the middle 50 percent of the data set. Trimming data does not typically eliminate half the data, but it does remove the most unlikely values, including any outliers, so that the results of an analysis are representative of an extremely conservative estimation.

Very frequently data will be missing not because it was removed from the analysis, but because it could not be collected in the first place. When you have a list of subjects, and only bits of data on some of them, then this can be problematic, since simply using the data you have available (known as pairwise deletion) will create distortions in the relationships between variables. In other words, some variables will incorporate subjects that others simply do not have, creating a discrepancy in the sample. The method of responding to missing data that will maintain the highest degree of validity is to use a listwise deletion, thereby eliminating that particular subject from the sample entirely. When dealing with data that requires unwavering precision, such as a scientific study, listwise deletion is the preferred method, but in applied analysis wherein missing data can be quite common, and one must weigh the risks and benefits working with uncertainty, data replacements are frequently implemented. Sometimes the missing value is replaced with the mean of the data set, and sometimes a value is randomly selected from a set of values of subjects with similar values of known data, but most commonly a value will be used that is consistent with the model built from existing data. In other words, whatever value the model would predict based on the data that is known, that predicted value is used in place of the missing data.

When assessing the relationship between variables, the potential exists that one or more of the variables being treated as independent actually have a degree of dependence either on other variables in the model, or on itself over time. When a variable is autocorrelated, it means that patterns exist within the data over time, and that the value of that variable at any given point is at least partially a function of its value at some previous point in time. This can be tested for using the Durbin-Watson coefficient, causing an overestimation of the strength of association given that self-perpetuating patterns exist. When several variables being treated as independent are actually at least partially dependent on each other, this is known as multicollinearity, which is tested for using a variance inflation factor, testing how much the correlation is inflated over the association that would exist if no association existed between the predictor variables. Multicollinearity can often be resolved by increasing sample size to eliminate the potential for multicollinearity resulting from sampling probability. If this does not resolve the issue, and the multicollinearity is strong enough, then removing one of the correlated predictor variables may be an option if the covariance is strong enough, or the two may be combined into a single covariate composite that isolates the idiosyncratic variability of each.

One of the most common problems that occur is one in which analytics is not related to the data itself, but human interpretation of the data: the confusion between correlation and causation. This is a very basic concept of all research, yet far too often it is a concept that is ignored. Correlation means that the value of one variable is useful in estimating the value of another variable, but that does not mean that one variable causes the other. The typical example used is one of ice cream sales and polio rates, which vary together as both are influenced by the temperature. A more modern example comes from the econometric research performed by Reinhart and Rogoff who demonstrated that there was a correlation between the amount of government debt and the economic growth of a nation. In their recommendations, they state that higher rates of government debt causes slowed economic growth, but had they performed the proper data diagnostics they would have saved themselves an embarrassing mistake. A simple way of testing whether causation is true is to use chronological analysis. In the case of Reinhart and Rogoff, had they noticed that slowed economic growth came before increased government debt, they could have avoided a very public error. Another way that they could have tested the validity of their conclusion was to reverse the order, as post-WWII UK experienced a decrease in the level of their debt but did not experience an increase in their economic growth rates. Frequently, information about the timing of the data will not be available, in which case analysts must be careful to draw only the conclusions that the data provides: that one variable can be used as a predictor of another, but does not necessarily cause the other to occur. The correlation can still be a valuable tool for predictive analytics (discussed in more detail in part II), but conclusions cannot be drawn in operational analytics (discussed in part III).

The final diagnostic with which this chapter will be concerned is regarding the manner of data collection, which must produce results that are both reliable and valid. To be reliable means that the collection of data using the same methodology will produce consistent, reproducible results, while validity refers to whether the data being collected is providing information about the correct thing. Think of it like target shooting: if your shot is reliable it will create a tight cluster on the target, but if it is not valid, you will be missing the bullseye. If it is valid but not reliable, some will hit the bullseye but only sporadically. To collect useful data, the methodology must produce results that are both reliable and valid. Reliability can be tested in various ways, such as performing the same test multiple times to ensure that the results are consistent. Issuing different versions of the same test to see whether they produce the same results, or different items measuring the same thing within a single test, can help to diagnose errors in the measurement or collection of data. Having multiple people separately interpret the same results is an easy way to test whether the interpretation of the results is reliable, but the most accurate method, if possible, is to apply the models derived from the data to historical data as proof that they properly predict what the outcome would have been using data prior to its occurrence.

Validity can be a bit trickier to diagnose, because there is never a 100 percent guarantee that any analysis is perfectly valid. Just as correlation does not necessarily equate causation, data that are useful does not necessarily mean it is valid, but as long as it performs as necessary, then it has functional validity. For all the research that is done that provides evidence supporting a theory, it only takes one study of enough significance to prove it all wrong. In order to minimize the possibility of this occurring, or at least to produce data that has functional validity, controls must be taken during data collection to eliminate threats to validity, of which there are three broad categories: external, internal, and construct (see tables 4.1 and 4.2).

The threat to construct validity refers to the possibility that the definitions and variables measurements being utilized to collect data are insufficient.

Specific characteristics which data sometimes exhibits can result in incorrect conclusions about descriptive analytics, and many tests in both predictive and operations analytics rely on these characteristics

Table 4.1 Threats to external validity

Type of Threat	Impact of Threat
History	Changes in the participants' environment over time influence the results.
Maturation	Changes in the participants themselves over time influence the results.
Regression	Extreme scores regress toward the mean over time as such extreme values can be difficult to maintain and behaviors seek equilibrium.
Selection	Bias in selection of participants.
Mortality	Participants fail to provide full data.
Diffusion of Treatment	Participants communicate with each other about the research, allowing them to change their response to it.
Compensatory	When only one participant receives intervention, participants may compare notes on impact and discover the cause of differentials in experiences.
Testing	Participant becomes familiar with the treatment and adjusts their response.
Instrumentation	Change of instrumentation between pre-test and post-test.

Table 4.2 Threats to internal validity

Type of Threat	Impact of Threat
Selection/Treatment Interaction	Participants that do not exhibit some intrinsic characteristic may not respond the same way.
Setting/Treatment Interaction	Participants may not respond the same way in a different setting.
History/Treatment Interaction	Participants may not respond the same way over time.

being eliminated, or at least mitigated. It is not enough to guess that the assumptions necessary to yield useful data are being maintained. It must be an ongoing process to diagnose and resolve violations of these assumptions, otherwise decisions will be made on incorrect analytical conclusions, resulting in harm or even total failure of operations.

It is important to note that many of the tests and calculations included in this chapter come with some problems. First, they tend to lack statistical significance with smaller sample sizes, which tends to be true of any statistical tests when n is small enough, but tests for the violation of assumptions tend to be particularly prone to lacking robustness with smaller samples. They also do not explain how much of a violation in the assumption has occurred. So, as with all analyses, these tests must be taken within context and used to supplement, rather than replace other forms of analysis.

CHAPTER 5

Challenges and Limitations

No field of study has limitless potential, or is without its challenges, and understanding what cannot be accomplished is just as important, potentially more-so, than knowing what can be accomplished. Each part of this book, then, will include a chapter briefly describing the challenges and limitations associated with the respective types of analytics being described. This is to help define exactly when it is appropriate to use these methods, when it is not, and what one might expect from their implementation.

Descriptive analytics are extremely limited in functionality, as they do only as the name suggests: they describe things. This is tremendously helpful in taking any type of measurement and in making comparisons and, while knowing with certainty exactly what has happened in the past and what is happening now is of critical importance to any operations at all, the range of things which can be accomplished with this information is pretty narrow. Descriptions do not allow a person to make accurate predictions about what is going to happen, and without a wide range of information about the context in which the descriptions are measured they do little to help make decisions. Much of the usefulness of utilizing descriptive analytics comes from incorporating them into other types of analytics. The contents of part I form the foundations of parts II and III—both predictive and operational analytics utilize the information derived from performing descriptive analytics in various ways, developing models that use descriptive analytics to infer other types of information. So, while the functions of descriptive analytics are few by themselves, when utilized properly they can be used in other analytics that would be impossible to perform without them. This, alone, makes them immensely important to any analysis that can ever be performed.

One of the most challenging parts of performing any analysis is the collection of data. First, there must be a way to measure the thing being studied which, in itself, is not always a simple process. Measuring the skills or knowledge that particular people in the opposition have is not something that can always be easily done even with the cooperation of the individuals. First, a method of measurement must be established, ensuring that the proper thing is being measured free of influences by extraneous variables which might invalidate the measurement being performed. Once a scale and method of measurement has been established, then actually performing the measurements to collect data can also be quite tricky, as some method of observing or otherwise evaluating the thing being studied is required, necessitating the development of observational tools (e.g., satellites) or, perhaps, requiring one to incur greater risks associated with contact. In behavioral research, when people know they are being observed, their behavior changes, invalidating the data being collected, which is known as the observation effect. People become more aware of their own behaviors, hide things which they do not want others to know, or even present misinformation intended to mislead, and so informed consent is not generally pursued under circumstances of conflict. Even when the data collection is performed properly, there must be something to measure. If the thing being measured has only occurred once, then there is very little available for comparison which would allow for an understanding of the cause, the influences on the outcome, and their implications. One might argue that the less likely something is to be encountered, the less vital it is to understand it, but the information derived from such assessments may not only help to shape events, but might reveal breakthroughs, which can be utilized in other ways. Sometimes it becomes impossible to directly measure the thing being studied, in which case measuring things that are related can function as an indirect measurement from which the desired information can be inferred. It is common during conflict, for example, to measure various aspects of resource distribution, pricing, and consumption; as well as various behavioral traits in the organization and activities of the opposition, to infer what future plans will be and how quickly they intend to act upon those plans. Since their intentions cannot be directly observed, most of the time, it is possible to measure those intentions by measuring the resources that they must use to accomplish their goals, as well as measure the behaviors one would expect to see in preparing to set-out to accomplish those goals.

Once the data has all been collected, it can be difficult to interpret, as discussed in chapter 4, on data diagnostics. Mere correlation is often confused for causation, the significance or reliability of results are often overestimated, and once a significant correlation is found people often tend to be satisfied with the results and stop asking questions. If not, though—if a full effort is made to collect ample amounts of the right data and continue to ask the right questions even after analyzing it—then the analytics performed, which describe those things being studied, represent the first step in developing the most precise decisions and strategies available to anyone.

CHAPTER 6

Suggestions for Future Research

In addition to providing a chapter in each part, detailing the challenges and limitations associated with the respective analytics described in each part, there is also included a chapter briefly touching on various potential directions that future research in the field could take. This is by no means intended to be a fully comprehensive or exhaustive list of all potential research that can be done. In fact, the potential range of research that could be pursued is limited only to the collective imagination and intellect of those who try, which is far greater than what any individual can conceive. Instead, this is only intended to be a description of those things that seem to me to be natural implications of those methods already being utilized, which are described throughout the chapters of each part. Often they are things related to facing the challenges or overcoming the limitations of the existing available methods. Sometimes they will be the result of those methods already being utilized included through the chapters in that part, but which have yet to be pursued. Other times they are items that have already been addressed in theory, or in other fields, but have yet to be incorporated into strategy development. Nothing listed in these chapters suggesting directions of future research are without some existing reference point, though. It would be an amazing accomplishment to develop analytics that can predict how specific future technological innovations will be made, allowing us to speed-up, and even "leap-frog" technologies, but there is absolutely no basis for such an analysis—it is not implied by anything currently being done. More than anything, the suggestions made in these chapters are directions of research that are expansions of the topics in these chapters I would hope had already been pursued, or which I would have liked to pursue, myself.

Descriptive analytics, even now, tend to be a little on the simplistic side, making it difficult to encompass the myriad of traits a data set can have in just a few calculations. Calculations tend to be generalized specific traits, to capture the overall sense of what is happening but not the specific rates in which they occur. This is why graphs and charts tend to be very useful in illustrating things that are not easily communicated using equations, but graphs and charts do not provide the kind of precision necessary to be effective in high-stakes environments. Expanding on the base calculations, which are performed in descriptive statistics, would be very useful in this regard, increasing the variety of tools that are available. The same is true for methods used in data diagnostics. Although there are several treatments that can be applied to data to resolve assumptions of violations, each of them have pros and cons, so exploring a wider range of options may prove fruitful in improving the available responses to specific types of idiosyncrasies.

Even more important, however, is the issue of how to respond to cases of missing or incomplete data. In nearly every field and every function, decisions are made under a balance of acquiring sufficient data, and actually making a timely decision; the result of which is generally that decisions are made without all the information that would be required to ensure an optimal choice. Under dynamic environments such as in a combat zone, it is particularly difficult to acquire all the data required, and so the issue of missing or incomplete data is one that is frequently encountered. The implication of this is that among the most important directions for future research is in finding better responses to missing or incomplete data. This would not be a glamorous finding that makes headlines, but it would improve the efficacy of nearly all data analytics that would ever be performed, making it an utterly indispensable and vitally important discovery. This could be accomplished by developing new treatments for missing data, or by developing methods of analysis which yield useful information utilizing smaller sample sizes.

Treatment of missing or incomplete data is not the only way to help minimize the impact of that problem—another method is by minimizing the amount of missing or incomplete data being encountered in the first place. This would be accomplished by focusing on improved sampling and measurement methodologies, improving the manner in which data are collected. This field of work would have the additional benefit of improving the validity of research, as well. This can be done by developing improved direct methods of measuring those things being studied, but this can be very difficult when there are people intentionally attempting to keep those things secret, or who are trying to harm you. There are also various ways in which to measure these indirectly. For example, by using some of the predictive analytics discussed in part II, it is possible to identify which independent variable will predict the value of the dependent variable. If those independent variables are easier to study, then they can provide an improved method of generating information. As an example, the detection of materials used in nuclear weapons can be done without actually identifying the material, itself, but by searching for elevated levels of gamma radiation, which is emitted by those materials. During WWII, intelligence agents were able to identify undercover vessels not by directly observing anything related to the vessels, specifically, but by watching for jumps in the price of black market goods and then correlating those jumps with the landing of vessels in the same time period, greatly narrowing the range of possibilities. Improving sampling method may include identifying new methods to tap into the entire population being studied, rather than varying degrees of bias created by limitations in what an analyst can access, or in developing new ways to filter a sample to more accurately represent the issue being studied, eliminating any inappropriate participants. In any case, it is of utmost importance that those being studied are not aware of the research operations, because behavior changes when they become aware, invalidating any future results being collected—or, at least, changing the nature of the study to evaluate how behavior changes and inferring what that change might mean, and how the "lie" behavior might provide information, in itself.

Again, there is potentially limitless other options in what can be researched; everything you can possibly imagine, and everything you cannot imagine, too. To remain prudent, however, there must be a basis for the research being performed—something that implies that the results may yield useful information, or which can accomplish something tangible. The nature of this chapter, and the remaining two "Suggestions" chapters, then, is to point out those issues which seemed to be natural implications of the subject of each part. They are meant to inspire thought and encourage people to consider not just what has already been accomplished, but what can be accomplished with the right resources.

PART II

Predictive Analytics

Predictions do not involve a crystal ball that allows you to see into the future. In fact, time is merely incidental for most predictions because although the things being predicted typically occur within a given timeframe, the prediction is being made using probability estimates related to variables other than time. Predictive analytics function by calculating the probability that the expected value estimated in the model will be the same as the observed value one encounters in reality. As this probability increases, it is said that the predictive value of the model increases. This means that by using the values of independent variables, a model can be built that predicts the value of a dependent variable, and that greater predictive value means that a larger percentage of the variability in the dependent variable is accounted for in the model as an effect of variability in the independent variables. The probability of being wrong, or the amount of deviation from the expected value, results from the amount of error in the model which does not properly calculate 100 percent of the variability in the dependent variable. In basic calculus, this is represented using the core idea that $f(x) = y$. In other words, the value of y is a function of the value of x; every time you put a different number in for x, a unique y number will be calculated. In making predictions, one is attempting to determine what y value will be observed in the real world by figuring out which x values will influence it. Typically, to account for 100 percent of the variability of y, there will be multiple predictors, $f(x, z, a, b,$ etc.$) = xy + zy + ay + by$, wherein each independent variable ($x, z, a,$ and b) is calculated for its influence on the dependent variable (y).

In a way, the predictions made by performance psychics such as John Edwards, or by astrological horoscopes, or any other supernatural-type predictions function on the same basic principles but their predictive value is too low to be functionally useful. A stage psychic from the United States can make broad predictions using simple probabilities—they can look at simple data from the Social Security Administration that there is a high likelihood that someone in the audience is named John, James, Mary, or Patricia, just because those are the most common names for adults in the United States. The psychic could even guess that one of these people has a son whose name ends with an "n," such as Aiden, Jaden, or other variations on this name, since these names have been increasing in popularity over the past 20 years, and in 2011 36 percent of boys born were given a name ending in "n." Most people have someone in their extended family, either alive or dead, who either has some medical condition, and the vast majority of the population struggles with finances. These are things that simply apply to such a broad volume of the population that the probability of being correct in one's guesses are extremely high by addressing a large enough sample in the form of an audience. Once a single guess is correct, such as a medical history, then having a knowledge of the most common medical conditions can further validate for the audience that the person is psychic, guessing things like heart disease, diabetes, Alzheimer's, and other conditions, which are increasingly common with a fat and aging population. Still, these things do not have enough predictive power to make them useful—they work by knowing that at least one person in a large enough sample will be representative of these common statistics.

The same is true of astrological horoscopes—the descriptions they give are so broad, and vague, that they encompass a variety of different interpretations and ideas that the probability of the reader being included is extremely high. These predictions tend to be too broad to be useful, however, in the same way that guessing an estimated time of arrival is, "sometime between now and 10,000AD." While that prediction is most likely true, it is so broad that you cannot make decisions based on it, illustrating the core trade-off in predictive analytics: precision and accuracy. The more broad a prediction is, its accuracy will increase, meaning that it will be correct a greater percentage of the time. Greater precision means that the narrower range of values gives a more exact understanding of what to expect from the observation, but that the prediction will be wrong a greater percentage of the time.

It is common to think that all predictions are attempts to predict the future, but this is neither the mechanics nor the goal much of the time. The mechanics of predictions, as noted, are always calculations of probabilities related to independent variables. Sometimes time is one of those variables, such as when predicting the relative strength of a battalion at various time-based milestones after initial mobilization of resources. Other times, time is just incidental, such as when predicting resource consumption rates and budgeted resource allocations, which occur over time but are functions not of time, but of the number of people and the activities in which they are participating. The number of bullets fired, for example, has nothing to do with time other than events occur over time, but it is a function of the number of times a firing-range training is conducted, and the number of people participating in each training exercise. Other times, time has nothing to do with the matter, such as when predicting the location of something by using "red flag" incidences, which are observations of independent variables known to be predictors of the dependent variable, then using a pattern of observations to predict the location of the thing of interest. Specifically, this latter method is known as abductive reasoning.

There are three basic forms of reasoning: deductive, inductive, and abductive. Deductive reasoning is a top-down approach to logic, which utilizes abstractions of existing knowledge and theories to draw conclusions about some relevant, specific idea or event. Inductive logic is the bottom-up approach to logic, which utilizes observations to form an understanding of abstract relationships that can be applied to a variety of other, similar circumstances. Abductive reasoning differs from the other two in that it uses specific observations to draw conclusions about other specific ideas or events, in a "best fit" association between them. Each are used under unique situations, depending on what information is available and what one is trying to accomplish, and each works together to supplement the other like cogs in a wheel. A solid prediction will be logically supported by valid theory, the theory will be supported by the observations, and the observations will be consistent with each other. When these three things are in alignment, then it is believed that the line of reasoning is true, and the most profound test of this validity is prediction: When something has predictive value, then it is functionally useful, even if all that has been established is correlation, and not causation.

Part II of this book explores the matter of probability in detail, including how to build probability models for use in estimating likelihood of observing particular values. The content in chapter 7 will then form the basis for the following two chapters on correlation, and geospatial intelligence. These allow us to predict with great accuracy hidden truths—things that are not yet observed, but knowledge of which can be inferred from related observations. When a model is developed, which has strong predictive power, then not only is it possible to optimize one's response, but those independent variables that are determinants of that thing being predicted can frequently be managed to alter the outcome. When such predictive variables have been identified, should a method of manipulating those variables be found, it then becomes possible to indirectly manage not only one's own operations, but the operations of others, as well. This transforms predictive analytics into a method for developing strategies for proactively controlling the opposition, rather than simply responding to them.

CHAPTER 7

Probability Modeling

Probability is a field of mathematics that quantifies and models the likelihood of some event occurring, making it identical to statistics in various ways. In fact, the two fields quantitatively describe the same things, but in different ways, as many analytical models incorporate elements of both. In defining the exact nature of probability, there are two separate views that can be understood in terms already addressed in this book. Frequentists refer to the rate of observation of various potential values of a variable, while Bayesians work on the underlying assumption of a statistical distribution like a normal distribution, and infer likelihoods based on that distribution. Contrary to common belief, these are not contradictory views of probability—they are complementary. While frequentists utilize sampling methods and the analysis of descriptive data, with the implications based on that, Bayesians heavily rely on inferential analysis with known behaviors of observation and data distribution acting as the source of evidential analysis. A common example used to describe probability is the roulette wheel, which is a type of gambling in which a large wheel is spun and a ball falls onto one number/color combination among 38 possible outcomes while people bet on which will be observed. Frequentists would measure the probability through sampling and measuring the frequency of observation, while Bayesians would model the apparent probability using a number of possible outcomes, each supplementing the other. Just as comprehensive statistical analysis incorporates both these approaches, so too does probability modeling. Despite these similarities, there are some profound differences between statistics and probability which warrant their being labeled as two separate, distinct fields. First, statistics is a broader field that incorporates not just probability, but many other types of analysis, while probability focuses on a more narrow range of topics. Probability forms the basis for much statistical analysis, but statistics goes beyond the parameters of probability analysis. The differences in analytical emphasis are no more obvious than in the notation used in probability modeling. Although all probability models can be expressed as statistical models, the opposite is not true, as not all statistics relies on pure probability. Despite the ability of probability to be expressed in statistical notation, to do so would often be unnecessarily complicated. The emphasis that probability has in analyzing likelihood, alone, has necessitated the development of some unique notation, which simplifies models that would otherwise be much larger than necessary.

Probability models look unique—filled with notation that a person is unlikely to have encountered unless they have studied probability—but being unfamiliar does not make them more difficult. The notation of probability merely simplifies a variety of things, which have either already been discussed in this book, or are considered common knowledge. For example, refer to this equation:

$P(A) \in [0,1]$

This simply means that the probability that A will happen is between 0 and 1; that $P(A)$ is a subset of the range 0–100 percent. Since all probability exists between 0 and 1, clearly the probability of A will also fit within these parameters, and that odd-looking symbol that somewhat resembles a drunken "E" merely states that the variable value $P(A)$ fits within the total range of possible values that follows it. If that symbol were facing the other direction, then the meaning would be reversed, as well; that $P(A)$ encompasses some value—that the other value is a subset of $P(A)$.

A more common type of calculation is the multiplication of decimals, which is basic arithmetic. Since all probabilities function as decimals between 0 and 1, one might expect to regularly see equations like 0.5 × 0.75, for which the answer would be 0.375. In probability, this would calculate the likelihood of encountering variables A and B, the former having a 50 percent probability of occurring on its own, and the latter having a 75 percent probability, while the probability of encountering them both is only 37.5 percent. In probability notation, this familiar model would look like this:

$$P(A \cap B)$$

If the probability of events is mutually exclusive, then the upside-down U would be flipped vertically, and rather than multiplying it would be necessary to add the respective probabilities of A and B.

Probability models are best visually represented through the use of Venn diagrams, as in figure 7.1:

The entire set of all values, S, has a probability of 1, since all possible values are within it. If sets P and R did not overlap—if they were independent of each other—then the calculations described earlier in this chapter would be used to determine the probability of encountering either exclusively (sum probabilities), or together (multiply probabilities). More formally, however, and particularly since P and R overlap, the equation used to determine the probability of choosing from set P or from set R is $P(A) + P(B) - P(A$ and $B)$, all of which can be entirely encompassed in the equation:

$$P(A \cup B)$$

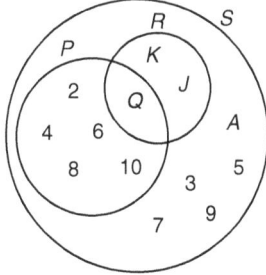

Figure 7.1 Venn diagram

Note that the probabilities of sets P and R are added together, and then the overlap between them subtracted so that Q on the Venn diagram is not counted twice.

It is also common to use conditional probability in planning. Conditional probability occurs when calculating the probability of something occurring given that some other event occurs. Using the Venn diagram in Figure 7.1, one might want to know what the probability of observing set R is, and then the probability of observing value J. In such a case, it would be said that one is calculating the probability of J, given R, notated as

$$P(A|B)$$

which is the same as saying

$$\frac{P(J \cap R)}{P(R)}$$

Conditional probabilities such as this are incredibly useful in calculating decision trees, asset allocation models, and other comparative models used in decision-making, discussed in greater detail in part III, especially chapter 15 on "Asset Management."

Once one becomes familiar with the notation used in probability, the actual details of the calculations are no different than any other equation one might encounter. Probability models serve the same purpose as any other form of analytical model, which was discussed in greater detail in chapter 2, and their design often becomes apparent when graphically illustrating them. Even calculating the probability of observing specific values can be modeled using a normal distribution, as discussed in chapter 1 on "Descriptive Satistics." For example, mathematician Lewis Fry Richardson demonstrated that the sizes of conflicts exhibit the traits of a normal distribution, wherein the vast majority of them will be of moderate size, while scarce few are extremely large or small, thereby providing information allowing for predictions on the size of an impending conflict. What makes probability unique is its exclusive emphasis on describing uncertainty, and the idiosyncratic manner in which notation and models have evolved over time from this emphasis.

The nature of probability was briefly touched-on in chapter 1 on "Descriptive Statistics," when discussing the traits of a normal distribution. The same distribution is used in probability, but is called a Gaussian distribution, named after Carl Gauss, who contributed greatly to our understanding of probability distributions, though some of his contributions are debatably attributable to others. As noted in chapter 1, under a normal distribution, a known percentage of the total values will fall within specified standard deviations from the mean; the probability of encountering a particular value can be modeled using the z-score of that value. Normal distributions are particularly important in probability as a result of the central limit theorem, which states that the underlying populations being studied are generally normally distributed, and that a sufficiently large sample will also fall within a normal distribution. In combination with the law of large numbers, stating that as a sample increases in size that its distribution will more greatly represent the population distribution, it becomes possible to generate a standardized approach to probability even without sampling—thereby providing the basis for Bayesian probability. The probability models associated with a normal distribution are used not only to determine the probability of encountering a particular value, but also to estimate the minimum and/or maximum values that are likely to be encountered, which values are considered outliers, the range of likely values for which one should prepare, and so forth. In fact, all statistical tests rely on the fundamental probability models utilized in a normal distribution in order to determine the critical value of the test and calculate whether the data are statistically significant or not; p-values commonly being 0.05, 0.01, and so forth, as discussed in chapter 1.

Normal distribution is not the only type of probability distribution, though. There are, of course, variations on the normal distribution which include deviations of skew, kurtosis, or even multimodal distributions, but in probability there are several more distinct types of distributions. Poisson distributions, for example, can be particularly useful in strategic applications. Poisson distributions, pronounced pwa-sohn, calculate the probability of observing specific values at a particular rate over time or over a geographic space. The classic example of a Poisson analysis is related to military operations, proposing an analysis of the rate at which Prussian cavalrymen were killed by being kicked by a horse, using the probability calculation:

$$p(k) = r \times k / (k!)(\varepsilon \times r)$$

Simply put, in the example, a Poisson distribution would calculate the probability of observing any specified number of kicking deaths in a given year. This type of analysis is extremely common in some fields, such as life insurance, which calculates the likelihood of death or the probability of paying claims on x number of deaths in a given year. These distributions are used to calculate entropy and radioactive decay, both highly relevant to traditional forms of combat. They have immense potential is planning resource needs during combat, contributing to estimates of the rate of attacks or encounters during a war, or even calculating the probability of a conflict occurring at any given point. This can also be modified to estimate probabilities over a geographic area, a concept that is discussed more greatly in chapter 9.

Related to the Poisson distribution is the binomial distribution. Binomial distributions specifically focus on the outcome of trials in which there are only two possible outcomes. Those outcomes can go by different names, but the mechanics are always the same; yes/no, win/lose, success/fail, on/off, 0/1, and so on. A Poisson binomial distribution calculates the probability of observing x number of positive outcomes from these trials, when the probability of each outcome is not necessarily evenly distributed. A special case of that, the ordinary binomial distribution, is distinct in that it assumes even distribution of probabilities in each trial. An even more narrow binomial distribution is known as a Bernoulli trial, wherein $n = 1$. In other words, Bernoulli trials calculate the probability of observing x number of specified outcome with equal probability given a single sample. Another way to look at it is that a binomial distribution is the distribution of n number of independent Bernoulli trials. These analyses are common in simulations, as well as in goal analysis.

In encountering a case of calculating the probability of outcome x given k possible outcomes, rather than just 2, the model known as a multinomial distribution is a generalized variation on the binomial distribution. Whereas a Bernoulli trial is an $n = 1$ variation on a binomial distribution, a categorical distribution is an $n = 1$ variation on a multinomial distribution. The core calculation of a multinomial distribution helps to illustrate the relationship between the different variations on the basic Poisson distribution:

$$p = \frac{n!}{n_1! n_2! n_3!} p_1^{n_1} p_2^{n_2} p_3^{n_3}$$

Note that this can be altered as necessary to include as many possible outcomes, each with their own respective probability, or sample sizes as required for any calculation. The most common example is the use of dice, wherein a simple multinomial calculation determines the probability of encountering a particular value given 6 possible outcomes, using x number of throws.

One of the most common types of distributions utilized in probability is the gamma distribution. The gamma distribution is another broad-based description of probability density distributions which includes some of the most commonly utilized distributions such as the exponential distribution and the chi-squared distribution, the latter of which is discussed in an applied context in chapter 3 on "Comparative Assessments." Gamma distributions evaluate the intervals between observations, and are frequently used to model predictions of intermittent frequency or duration over time or across a geographic area. The amount of time between attacks, for example, or the amount of space between theater operations can be modeled using estimates of the changing rate of probability of encountering an interval of particular size.

Probability distributions form the underlying foundation upon which analytics can be performed, which calculate the probability of a particular outcome. The most common of these is called logistic regression, in which probability, p, is calculated as:

$$\frac{e^{mx+b}}{1+e^{mx+b}}$$

Note that the term e^{mx+b} models a linear relationship of a predictor variable on probability as an exponent of Euler's constant, e. Logistic regression is used to determine the probability of a dependent variable occurring as the value of one or more independent variables change value. For example, the probability of a successful mission might be a function of resource availability, and experience, so that as each of those independent variables increase, the probability of mission success also increases. The strength of the relationship between these independent variables and changes to probability can be assessed using the correlation coefficient, r. This makes logistic regression extremely versatile and useful for estimating how likely something is, or will be, based on variables which are known. This helps to calculate exactly how certain you are of something based on variables other than direct observation, which is particularly useful under conditions of limited information.

The final method of probability modeling that will be covered in this chapter is known as Bayesian probability. Named after Thomas Bayes, Bayesian probability stems from Bayes Theorem:

$$P(A|B) = \frac{P(B|A)P(A)}{P(B)}$$

Bayesian probability is, by nature, subjective, and therefore relies entirely on conditional probabilities, wherein the probability of some outcome can be modeled, conditional of the outcome of some other variable. The models remain abstract, rather than empirical, but provide immense value in deriving inference, comparative probabilities, relationships between variables, and analyzing adaptation to change.

While modeling probability, by itself, can certainly help to describe and predict locations, actions, events, timing, and so forth, it cannot actually do anything to alter these things. Just knowing the probability of an occurrence in enough to have an immense impact, such as the use of predictive policing, wherein simply knowing where crime will occur allows for the distribution of law enforcement resources to that area at times the crimes will occur in order to prevent them from happening, at all—something also possible with combat (discussed more greatly in chapter 9, on "Geospatial Intelligence"). This demonstrates something even more profound than prediction: manipulation. The volume of law enforcement resources available in an area is a critical variable to the probability of a crime occurring, so by allocating those resources properly the probability of criminal activity goes down, resulting in real reduction to the amount of crime actually committed. Once the independent variables that influence probability are identified and modeled, it is entirely within one's capabilities to alter the value of those variables proactively in order to change the outcome of events to one's own favor—either making desirable conditions more likely, or undesirable conditions less likely. In this way, probability modeling becomes something of a roadmap, defining exactly the actions one needs to take in order to alter a course of events through cause and effect, thereby controlling fate.

CHAPTER 8

Correlative and Regression Analyses

Everything happens for a reason. That does not mean that some supernatural force or metaphysical entity is governing the events in our lives in some predestined game to which we respond, as if each person is the focal point of their own narcissistic universe, custom-tailored to guide us through existence. It simply means that there is something that caused each event or action to occur, and that everyone is subject to them. Whether or not your organization wins in some conflict or competition has nothing to do with gods or fates but, rather, is entirely dependent on the performance of the organization which is, in itself, dependent on the performance of the organization's leadership. Whether you get laid-off of your job depends on the performance of your company and the skills demanded in the current labor market. It is not the fault of fate if companies do not need someone with your skills, it is up to you to ensure that you keep your skills up to date with the changing needs of companies. If a company files bankruptcy, it was not a god that caused this to happen; in fact, it is possible to predict whether a company will file bankruptcy with 90 percent accuracy using an analysis called Altman's Z-score, which utilizes a predefined series of weighted variables related to the performance of the organization to make that prediction. When someone is injured on the battlefield but survive, too often they thank their god, while the real reason they survived was that a doctor spent many years learning to heal war wounds, and the injured person's command was able to mobilize the resources necessary to make that doctor available to them. In such a situation, the likelihood of your survival depends on the volume of medical resources available, how long it takes to mobilize them, the experience and skills of the medical staff, the location and type of wound, and a several other, less influential variables. With the right kind of data, it would be possible to predict with great accuracy the probability of your survival, just as it is possible to predict which side will win in a conflict, or something simple like the amount of resources that will be consumed in sustaining normal daily operations. In each of these examples, the outcome is the direct result of multiple, measurable variables, allowing for an accurate prediction of that outcome.

This holds true of all thing, even including the decisions each person makes. The brain is basically a chemical computer, composed of electrical impulses and chemicals called neurotransmitters. Each person is born with instinctual responses such as suckling, as well as the basic framework for learning, behavior, and emotional responses, which functions something like the human equivalent of computer firmware. The majority of our decisions and actions are based on things we learn—not just information we can use, but cultural traits and beliefs we inherit, fears and hopes that we develop in response to our experiences, behaviors and habits that we are conditioned to perform, and so forth. Thanks to simple correlation, it is possible to predict with reasonable accuracy a wide variety of things about a person purely based on where they grew up, such as their religion, the kind of foods they eat, the kind of housing they live in, the language they speak, and behaviors they exhibit, the dynamics they have with the family and employer, and more. Certainly, rocks have no ability to make conscious decisions, so the movement and physics of the planets and stars are also readily predictable. As Isaac Newton stated, "I can calculate the movement of the stars, but not the madness of men," but since the time of Newton we have developed methods to calculate that madness. Psychology, itself, is little more than applied physics, since the chemical–electrical reaction that is our mind also follows the laws of physics and organic chemistry. As a result of our ability to predict both the physical world and the human one, theoretically a single mathematical equation could have been developed from the moment the universe was born that would predict exactly everything that would ever happen until the universe ends. Everything happens for a reason—that reason being that each action and decision is the product of everything that led up to that point, meaning that everything was put on a predestined path at the moment of the big bang, and we need only learn to read that path as one would a map in order to tap into the immense potential that would come with seeing the future.

Making this even more useful, it is possible to make accurate predictions not just based on measurements of causation, but also that of correlation. Not-causative correlation occurs when the outcome of one thing can be predicted using measurements of another, despite the fact that one does not actually cause the other. Generally, these cases are caused by some third variable, which influences both; although specific conflict may be lasting longer and has a higher rate of failed missions, making it appear as though the failed missions are causing the extended duration of the conflict, it could actually be that the neglect of psychological operations is both causing individual missions to fail while simultaneously making it difficult to end the conflict. These non-causative correlations can be misleading, but once they are identified for what they are, they can also provide valuable insight. Even if there is no causation, a sound correlation will still have the ability to predict outcomes. Particularly when the true cause isn't known, or cannot be easily measured, using noncausative correlations can be immensely beneficial. There are also times where two variables are correlated, and causation is suspected but it cannot be determined which variable is causing the other. In cases such as these, the variables must be treated as noncausational until more research is done to determine what the causation is.

Correlations are measured by determining how much the value of the dependent variable changes in response to changes in the value of one or more independent variables. This is measured on a scale of 0 to 1, wherein 0 means that 0 percent of the variance of the dependent variable is associated with variance in the independent variable(s), and 1 means that 100 percent of the variance is accounted for by variance in the independent variable. A value of 1 means that the variables are perfectly correlated; as one variable changes by 1 percent, the other changes by 1 percent as well. In these cases, it may be prudent to ask whether the things being measured are actually distinct from each other, or whether you are measuring the same thing twice. The value can also be positive or negative; a positive correlation indicates that as one variable increases, the other does, as well. A negative correlation, calculated as a negative value, means that as the value of the independent value increases, the value of the dependent value decreases, indicating an inverse relationship. If a correlation has a value of 0.8, regardless of whether it is positive or negative, it means 80% of the variance of the dependent variable is explained, while 20 percent of the variation is yet unexplained; there are still variables that influence the outcome not yet included in the analysis.

Any of the comparative assessments discussed in chapter 3 can be used to measure correlation, assuming the strength of the correlation is strong enough to be significant. Since comparative assessments measure the amount of difference between different groups, they can also be useful as a method of gaining insight into correlation. The critical thing to remember when pursuing this is to ensure that the test being used is appropriate for the number and types of variables being evaluated. Even when performing these tests, though, it is generally appropriate to also perform one of several common types of calculations that are dedicated to analyzing correlative associations.

The most basic type of analysis is simply called the correlation coefficient, or Pearson's r, which calculates the amount of correlation between two variables. Graphically, this is typically represented as a scatterplot, with a line drawn in the middle of them that represents the average of the relationship. It is calculated as follows:

$$r_{xy} = \frac{n\sum xy - n\sum x \sum y}{\sqrt{\left[n\sum x^2 - \left(\sum x\right)^2\right]\left[n\sum y^2 - \left(\sum y\right)^2\right]}}$$

If this looks somewhat familiar, that is because, like ANOVA, it relies on several sums of squares calculations. The correlation, itself, measures only how strong the relationship between variables is, though, by assessing the comparative rates of variability between them.

What makes correlated variables even more valuable is the fact that they have predictive power. The simplest way to tap into that predictive potential is through something called linear regression, which can be best understood in terms of very basic equation from introductory algebra: $y = mx + b$. While correlation refers to the amount of relationship there is between two or more variables, regression is the equation that models the average relationship—it is the line on the scatterplot that illustrates the average manner in which one variable influences the value of another. As a result, a regression line allows you to take any value for the independent variable and determine what the outcome for the dependent variable will be or, vice versa, determine what value of the independent variable must be observed to achieve the desired outcome for the dependent variable. The equation $y = mx + b$ is the one used to graph straight lines; m determines how steep the slope is and what direction it goes, while b tells you the value for x at which the line will cross the y-intercept. In other words, if you encounter the equation $y = 1/2x + 5$, it means that when $x = 0$, that $y = 5$, and that the slope increases in value along the y-axis (vertically) 1 unit for every 2 units it increases in value along the x-axis (horizontally); best remembered as "rise over run." The value m is simple to calculate, requiring only two points on the line, since the line does not have any curves that would alter the rate of correlation between the two variables: $m = (y_2 - y_1)/(x_2 - x_1)$. The y and x values refer to the coordinates of each point chosen on the line. This is all very basic algebra, and sets the foundations for regression analysis, though. As noted, linear regression is a straight-line type of correlative analysis, which allows for prediction using a linear model based in the equation $y = mx + b$. Linear regression models the correlation of two variables as a straight line, for example, demonstrating that as the value of an independent variable increases by 1 percent, the value of the dependent variable increases by 2 percent, on average. Of course, unless $r = 1$, each of the observed values will deviate somewhat from the regression model, so that the prediction is not perfect, but that the predicted value will fall within a given standard deviation from the model; and the r value decreases, showing that the correlation is weaker, the predictive value of regression diminishes. The stronger the correlation, the greater the predictive potential of the regression model.

The calculation of the slope of a regression line is as follows:

$$\frac{\sum xy - \left(\left(\sum x \sum y\right)/n\right)}{\sum x^2 - \left(\left(\sum x\right)^2/n\right)}$$

The x value at which the line crosses the y-axis is calculated as follows:

$$\frac{\sum y - b \sum x}{n}$$

Once these two things are calculated, the model has been formed, and predictions can be made. It is important to remember, though, that any prediction is only as good as the underlying model.

In order to develop a quality regression model, it will generally be necessary to use something more complex than simple, linear regression. As noted earlier in this chapter, unless your r value is 1, there is a percentage of variability in the value of the dependent variable still left unexplained by the value of the independent variable. In order to improve the model further, we move to multiple regression. Multiple regression is not so much more complex than simple linear regression—the only difference is that multiple regression incorporates more than one predictor variable into the model to account for a greater proportion of variability in the independent variable than a single predictor variable can predict on its own. To put it simply, multiple regression is basically nothing more than $y = mx_1 + mx_2 + \cdots mx_n + b$, wherein n refers to the number of predictor variables being utilized. This is how Altman's Z-score, mentioned briefly in the introduction to this chapter, functions; no variable included in the Edward Altman's equation could accurately predict whether a company would file bankruptcy on their own, so he used several different variables, each with just a little bit of predictive power, which had a significant amount more predictive potential when used in combination. To further refine this model, he then weighted each variable to alter the proportion of the total model that each contributed. A very basic weighted multiple regression can be as simple as: $y = 1.2(mx_1) + 1.4(mx_2) + b$, as the second predictor variable is given more emphasis than the first.

When performing multiple regression, though, it is important to remember that more predictor variables are not necessarily better. As more variables are added to the analysis, unless they are treated properly, they will actually harm your results. Not only does the increased number of predictor variables complicate efforts to derive useful information about the nature of actual correlations, but incorporating too many variables can cause distortions in the results caused by coincidences, covariances, and chaos. The most effective models are those which are minimalist—including only the bare minimum of variables and calculations. Still, if a legitimate correlation exists, it can improve the predictive power of the model when incorporated appropriately, which means exploring the exact role the variable plays and utilizing quantitative analysis to determine how one influences the other, perhaps experimenting to explore why.

If variables exist which, by their very nature, cannot be eliminated from the analysis, or should it become necessary to understand the relationship between several variables while keeping something else constant, then it is time to use a variation of the correlation analysis known as partial correlation. If, for example, the variable "command experience" influences probability of mission success, but the variable "regional awareness" not only functions as a predictor variable in its own right, but also influences "command experience," then it may not be possible to identify the correlation that "command experience" has until controlling the influence of "regional awareness." The base equation, or "first order" equation, for calculating partial correlation is as follows:

$$\frac{r_{xy} - r_{xz}r_{yz}}{\sqrt{(1 - r_{xz}^2)(1 - r_{yz}^2)}}$$

The r values in this are exactly what they seem: the r value resulting from the correlations between the values specified. With higher-order partial correlations, increased numbers of correlations must be accounted for in order to hold a single variable constant within the analysis.

Sometimes a dependent variable is correlated with itself—the value that it held at some point in time previous to the one being predicted. This occurs when there are patterns in the value of a variable over time, either cyclical trends or self-perpetuating influences, such as positive and negative feedback loops. Feedback loops occur when the result of a variable's value causes changes in that variable. For example, experience makes a person more likely to succeed in their mission, while mission success implies that they survived to take experience from their experience. The US Defense Advanced Research Projects Agency (DARPA) has commissioned research into patterns and trends in physical variables over time, including autocorrelations, under a project known as Biochronicity and Temporal Mechanisms Arising in Nature (BaTMAN). The goal of this project is to turn the analysis of biological mechanisms from a descriptive process to a predictive one.

Factor analysis and path analysis are two quantitative tools available, which are unique in a way that rather than attempting to describe and predict as other correlative analyses do, are more exploratory in nature. Factor analysis calculates the amount of covariance between unlike variables, and evaluates their potential to be used as a single, aggregate variable known as a factor. The goal of factor analysis is to make calculations more effective by reducing their complexity, and increasing their analytical strength. Path analysis, by contrast, evaluates correlations in an extended, linear model, showing multiple correlations in a single model. The goal is to describe the cause and effect of long series of variables, demonstrating chain reactions wherein changing a single independent variable causes not just a change in a dependent variable, but has secondary and tertiary effects as that dependent variable functions as an independent variable in other correlations.

In chapter 3, on "Comparative Assessments," several types of measurements of effect size were described. Recall that effect size refers to the strength of the calculation, such as how the strength of correlative analyses can be measured with Pearson's *r*. Although Pearson's

r is the standard test of correlation, it is also a measurement of effect size for simple correlation. When using an ANOVA or variations on the ANOVA as a form of correlative analysis, the effect size can be calculated using either partial eta squared, or omega squared, both of which rely on the sum of squares calculated in ANOVA. Eta squared is calculated as $SS_{effect}/(SS_{effect} + SS_{error})$, while omega squared is calculated as $(SS_{effect} - (df_{effect} * MS_{error}))/(MS_{error} + SS_{total})$. Partial eta squared measures the amount of variance in a sample. Notated η_p^2, partial eta squared is an improvement on the original eta squared that helps to correct for overestimations in the effect size of the original, which increases with smaller samples and with increasing volumes of independent variables. Omega squared, by contrast, measures the variance of the population, rather than the sample, and is notated ω^2.

When calculating the effect size of a multiple regression, the most commonly used equation is called Cohen's f^2, calculated as $r^2/1 - r^2$. Calculating effect size of an analysis is a critical part of any correlative analysis, and Cohen's f^2 is very versatile in its application to these types of analyses. This original form, however, is very broad, and evaluates only the effect size of the entire model. In order to isolate individual elements of a multiple regression so as to measure the effect size of specific independent variables within the model, a variation on Cohen's f^2 is used, calculated using the equation

$$\frac{r_{ab}^2 - r_a^2}{1 - r_{ab}^2}$$

This form of Cohen's f^2 allows for the increased flexibility to isolate individual independent variables, or combinations thereof. This is very useful for most correlative and regression analyses discussed in this chapter.

Correlation and regression have immense potential when applied appropriately, but it is critically important to understand their limitations. The value of these analyses is highly dependent on the data used, and tends to distort extremely easily when extraneous or faulty data are incorporated. Even with appropriate data, predictions utilizing regression will only be estimates, the accuracy of which will depend on the strength of the correlation, so reporting on the likely range of possible values of a prediction will be vital to making decisions appropriate to the data. As noted in the beginning of this chapter, correlation does not always mean causation, so unless there is solid evidence to demonstrate that one variable is directly influencing another. Such an assumption will remain unwarranted, and the relationship should be considered more of a covariance, than anything. Although these limitations seem wholly restrictive, the information that can be derived from these analyses is vital, and provides insight into things for which no accurate estimates were previously available. This can be particularly useful once a person learns to think in terms of matrices, wherein the value of some variable is thought of not as a singular, but as a range of possible values to which one must prepare a response. For example, though it may not be possible to predict the precise capabilities of the opposition, by using regression it will become possible to predict the minimum and maximum possible level of capability, as well as determine the range of what is most likely to be encountered in a given circumstance. While individual encounters may vary within a predictable range of values, the overall trends experienced over a series of encounters will adhere more strictly to the model given, allowing one to effectively predict the future.

CHAPTER 9

Geospatial Intelligence

Place a map of proper scale and orientation over a scatterplot and several things happen that allow for an entirely unique set of analytics, which will be referred to in this book as geospatial intelligence. More precisely, a map can be used as the basis of a scatterplot in which individual observations are plotted in a manner that defines both the location where the observation was made as well as the time the observation occurred. There are a variety of software programs available specifically designed to analyze this type of information already available not only for military use, but also for use by private companies. From this data, information can be derived using not just regression analysis, which was discussed in greater detail in chapter 8, but also a best-of-fit analysis using something called abductive reasoning. In addition, important patterns have been found in the analysis of the geographic movement of people and resources that allow for specific types of geographic information not only predictable, but can even be proactively managed to directly alter the actions that opposition forces perform. People, it turns out, natural follow many of the same laws and models found in nature, such as those observed in animals, and even fundamental constants such as gravity. The success of any mission relies heavily on the ability to mobilize people and resources, and the logistical completion of the entire supply and value chain. In other words, if things are not where they need to be when they need to be there, everything fails.

From the perspective of strategic quantitative analytics, the manner in which the process of moving things around within a geographic domain can be accurately evaluated and manipulated; where things are going, or where they are can be predicted and managed; and information about impending actions being planned by the opposition derived from the manner in which they organize the geographic distribution of the resources that are available to them.

Most of the current analytics and analytical computer programs associated with geospatial analytics rely on a process known as abductive reasoning, which essentially refers to any analysis that uses data observations to find the "best of fit" response to missing data. For example, it is possible to predict with great accuracy the location where a particular type of crime will occur, where gang violence will happen, where insurgent groups operate, where opposition groups store their resources, and more. This is all done using methods that have been utilized for decades by private companies like hotel chains and fast food restaurants, when deciding the optimal location and size of a new outlet. This wide ranges of applications all have the same basic analytical mechanism, and simply use different variables within it. It all starts with the observation of some "red flag" event—the occurrence of something that has a strong correlation with an outcome of interest. It is common for government agencies, for example, to scan Internet sources, such as social networking and email, for words associated with violence. The Department of Homeland Security, which is the US federal agency tasked with domestic defense against terrorism, issued a list of the words they scan for under a Freedom of Information Act (FOIA) request by the Electronic Privacy Information Center (EPIC), including a variety of words associated with border violence, terrorism, cybersecurity, and more. These words are commonly used and are not typically associated with anything malicious, but there is a minimum level of concentration or frequency in which they can be observed at which point the elevated level of their use does, in fact, correlate with outbreaks of violence. This is essentially how the US National Security Agency's (NSA) PRISM surveillance program functioned, as does Epidemiological Modeling of the Evolution of Messages (E-MEME), Worldwide Integrated Crisis Early Warning

System (W-ICEWS), Riftland, and more. These programs, and others like them, utilize something called metadata, which means they are not concerned with the data itself, but data about the data. In other words, if one of these programs scanned one of your social networking pages, it would not evaluate all the words on your page, but it would contribute to data related to the number of the times a particular word was observed within a given geographic area. If a given unit of geography has an elevated frequency of these words, it is considered a red flag, because there is a very low probability that such words would be used so frequently in normal conversation—in a normal distribution, that such words are used so commonly at any given time without being associated with impending violence would be considered an outlier. One red flag is not enough to warrant further investigation, but when enough red flag events are observed within a given geographic area, and within a given timeframe, then it is considered cause for concern, and a manual investigation is triggered. Each source of data is associated with a different range of possible sources, and with several observations of predictor data it becomes possible to analyze within a particular radius the source of the events. SCARE software (Spatio-Cultural Adbuctive Reasoning Engine), for example, can predict within 700 meters (less than half a mile), the location of munitions dumps and weapons caches by using a best-of-fit analysis of these red flag events. Crush software (Criminal Reduction Utilizing Statistical History) was able to reduce crime by 30 percent, and violent crime by 15 percent, over the course of less than 10 years in areas where it has been utilized by law enforcement.

Most applications of geospatial intelligence rely on this sort of abductive reasoning. Hotel chains, for example, can determine the number of rooms they need to build in order to maximize profits by assessing population size, population density, tourism and travel volume, volume of attendance at annual conventions and exhibitions, and other data. Some fast food chains decide where to open new stores by looking at population density, demographic data, competition and other stores of the same brand in the area, and so forth; and they can accurately predict the volume of customers they will get over time, and how long it will take to break-even on the initial cost of opening the store. In making geographic predictions, the location of most anything can be made by identifying correlated variables and effectively collecting data on the observation of those variables.

Often geospatial analytics are not used to identify the location of something, though. Sometimes it is necessary to identify movement, or changes in the density of observations across an area. In these cases, it is possible to use regression analysis not so different from that discussed in chapter 8. Once again, it starts with identifying historical trends in the observation of red flag events, such as witnesses reporting seeing a particular person, group, or activity; or the use of a credit card to purchase fuel while tracking someone across long distances. Historical locations, alone, are enough to make predictions regarding current location within a given radius, though the size of that radius expands both over time, as well as when making predictions of movement over larger areas. Much of this error can be reduced, however, by including the time of observation, since that information can be used with estimates of the rate of movement to determine how far from the last observed location one is likely to be.

This type of regression analysis can also be used to evaluate strategic positioning over spatial areas. By collecting data of the density of observations of opposition forces, generally through sampling, it is possible to do a regression analysis, including testing for heteroskedasticity, and evaluating the rate at which density transforms over an area. Doing this, it is possible to evaluate the oppositions' front lines for weak or strong points to either be tested or avoided, respectively. It is also possible for companies to determine the regional strength of their competitors within various markets, such as the iron grip that Wal-Mart has on much of the US Midwest, while other chain stores such as Meijer's cannot compete in these regions, but remain strong in the Great Lakes area where Wal-Marts have lower density. By evaluating the density of Wal-Marts over an area, their competitive strength in that area can be compared over a larger region. This same premise can be used to determine the strength of insurgent forces by evaluating the frequency of their operations in specific villages.

Perhaps, even more profound than the ability to track and predict locations, movement, and density, is that there are patterns that exist in these things. An abductive analysis, by itself, will be effective is identifying the location of some event, but in-depth analysis of the geospatial data that has been collected many years have yielded results that show that there are underlying patterns to the manner in which people organize and distribute themselves and their resources. For example, research carried out by a team led by Dr. Brantingham of UCLA has demonstrated conclusively that gang, and some insurgent activity follow the same models of territory, movement, and conflict as those which apply to the interaction of predators and prey in standard ecology. Called Lotka-Volterra models, named after the researchers who developed them, these equations demonstrate predictable patterns in the rise and fall of population levels of predator and prey, the expansion and movement of their territories, and their rate of interactions. About 58.8 percent of observed encounters will occur within 1/5 of a mile of a territory boundary, 87.5 percent within 2/5 of a mile, and 99.8 percent within a full mile. Although many have an inflated sense of uniqueness, people exhibit many of the same behaviors observed among even humble animals such as mice, and observations of gang conflict held to within a 5% error margin of those predicted by the standard model. This is then used to determine where law enforcement resources should be distributed in order to prevent violent incidences from occurring, since the location of these incidences can be predicted in advance—something that is also possible during warfare, private competition, or a variety of other circumstances. Since people adhere to the same models and patterns exhibited in other fields of study, there is an extensive amount that can be derived from a variety of studies to define the actions that people will take. In addition, the Tracy-Widom Distribution is a specific type of skewed distribution common in a variety of scientific fields which models the sustainability of interactions between multiple groups, wherein interaction frequencies above a given level will cause one group to be overwhelmed by the spread of another.

During WWII, prior to D-Day at Normandy, Allied forces executed a complex, multi-phase strategy called Operation Bodyguard. This included several smaller operations, each focused on creating diversions. Each operation included as a part of Operation Bodyguard created artificial military operations by utilizing props, fake build-ups and movements, methods of tricking radars into reporting planes and ships in locations where there were none, and so forth. By creating false data for the Nazi military to analyze, they distributed their resources in a manner that addressed each of the fake missions, thereby decreasing the resource density at Normandy, creating a weak point for Allied forces to storm the beaches on D-Day. Although this is a rather simplistic example, it was successful in accomplishing the goal of diverting Nazi resources away from Normandy, and demonstrates how the movements and actions of opposition forces can be manipulated in predictable ways by altering the geospatial data, which is available. As discussed in chapter 7, once the predictor variables that influence the outcome of a particular action or event are identified, it is possible to control the outcome by altering the value of those predictor variables, allowing one to remotely control the operations of opposition forces.

Even Newton's models of universal gravitation can be utilized to develop strategies based on geospatial intelligence, since the natural behavior exhibited through the geographic distribution of people and resources follow many of the same phenomenon observed in the movement of gasses and rocks in space can also be seen on earth through something called economic gravity. Looking at the growth of any city, labor specialization, and trade attracts people and resources as increased production capabilities create a growth cycle wherein more people migrate toward the areas of higher population density in order to find customers for their own products, find a source of raw materials for their operations, find employment, find things in which to invest, and so forth. The ability of these cities to draw people and resources toward them is called economic gravity. This sounds simple enough, but the implications are far-reaching. For example, these basic economic forces transcend political borders, creating frictional interaction along national borders wherein the movement of people and resources between nations with contradictory goals causes points of contention. As mathematician Lewis Fry Richardson demonstrated in the early twentieth century, the propensity for war between two nations is partially a function of the length of the shared border between those two nations. Cities also have energy that push resources and people out away from them; once they get big enough, investors and businesses will move operations to more rural areas where it is cheaper to function, the rich will move away from the city to areas of lower population density where there is less crowding and all the negative things that come with crowding, and the capabilities that are associated with having resources and high-skilled individuals in an area are lost, creating pockets of poverty and blight. Since militaries do not actually produce anything on their own, maintaining military operations away from the home nation requires a lot of economic energy—the kind of energy that can be easily monitored by collecting data on supply lines and sources of capital funding. The role of international financiers in combat became a critical component of international relations associated with operations in Afghanistan and Iraq, as insurgent groups were largely funded by a handful of wealthy individuals living in nations inaccessible to US and UN forces under international law, but vital information about the role of these individuals and how to disrupt the distribution of that funding became available. The US Civil War was largely won based on differential in supply-chain capabilities, wherein the North had access to sophisticated rail networks, and attacked both rail and water freight systems in the South in order to starve the Southern troops of vital resources. Data regarding the shelf life of rations, the time it took to transport resources, and the trade of Southern cotton for military supplies all became critical aspects of the Northern strategy.

These natural patterns of human movement also allow one to derive information about the nature of conflict, and the intentions of particular groups across a region. Groups that operate without local production, or without attracting local people and resources will clearly have a strategic interest in the location strong enough to warrant the inflated resource consumption associated with remote operations, and without defined operational benefits as one would expect from legitimate operations, something covert must be taking place. Suicide bombers are obviously most likely to attack in densely populated areas, while planning and administration are done in remote villages or rural hideouts, but resources must still be sourced from urban areas, which have ample availability of recruits and supplies. Data about the purchase and movement of capital, resources, and people can provide detailed information about the functionality and organization of opposition forces, as well as accurate predictions about their planned future activities for which they are preparing. The movement of resources, when publicly announced, can send a very strong message, too. The mere presence of military forces in a region not only communicates intention, true of not, but creates a predictable response, as well. Each time the United States allocates resources to an Eastern European missile defense program, other nations in the area such as Russia and China respond with predictable actions, which attempt to discourage continued pursuit of that movement. During the Cold War, the respective space programs of the United States and USSR were intended to be demonstrations of the development of stable intercontinental ballistic missiles (ICBMs) that can carry nuclear warheads halfway around the world, which utilize the same rocket technology as that which send people into space. In 2014, North Korea attempted to send a message to the rest of the world by moving missiles into launch position, as though they were planning to launch them at South Korea, Japan, or the United States, in a type of confrontational diplomacy known as "saber rattling." China moved some of its own missiles to its southern border near Taiwan as a result of the conflict between the two nations (or two factions within one nation, depending on who you're talking to), in an approach similar to North Korea. All these example have something in common: they either collected or communicated military data intentionally to other nations by moving people and resources around to different locations. Lewis Fry Richardson, in his research on global conflict, showed that the rate of armament build-up is directly proportional to the amount of arms held by a given nation's rivals and the amount of grievance held toward that rival, and inversely proportional to the amount of arms that one already possesses. Arms races such as those experienced between the United States and USSR, or India and Pakistan are the result of these correlative relationships. The future volume and location of military resources can thus be predicted using data regarding the relationship between a nation and the rest of the world.

The patterns of interaction, conflict, and movement of military assets across a geographic region follow predictable patterns, in much the same manner as that of predators and prey described by the Lotka-Volterra models, as demonstrated in border and territory conflicts in the Middle East, and territory disputes in Southeast Asia. These patterns make opposition activity not only predictable, but also manageable. Even individual idiosyncratic activities can be predicted using either abductive analysis, or modified forms of regression analysis described in chapter 8. As so much is dependent on the geographic movements of interacting organizations, from their front-line operations to their high-level strategy, the ability to predict, respond to, and even alter these movements creates an immense advantage in any form of conflict. By understanding the logistics of organizational behavior, it becomes possible to prevent conflict from beginning, at all, by anticipating it and responding in a manner that limits the ability of those with opposition sentiment from organizing.

CHAPTER 10

Challenges and Limitations

Predictive analytics function entirely by using the value of predictive variables to determine the outcome of the variable being predicted. As a result, the ability of a model to make accurate predictions is dependent entirely on how many of the variables that influence the outcome being predicted have been identified, and whether each is being utilized properly within the model. Theoretically, all things can be predicted; every possible event and action are the result of variables that led up to that point and it will, in turn, be a factor that influences future events and actions. In the physical world, all things follow specific laws, allowing us to predict and even manipulate the evolution of plants and bacteria, the movement of the planets and asteroids, and even the weather. Throughout the twentieth and twenty-first centuries, it has been established that even the behaviors exhibited by people, as well as the emotions they feel and the ideas they have are all the result of a combination of two things: the genetics they receive from their parents, and the things to which they are exposed to throughout their lives. Religion is a geographically inherited trait; if you were born in the Middle East you would almost certainly be Muslim, while if you were born anywhere throughout the Americas you'd likely be Christian. The emotional response you have to any event is caused by a shift in the chemicals in your brain and throughout your body that is triggered by your genetic propensity to respond in particular ways, as well as the meaning you have learned to attach to specific events throughout your life. For example, the meaning that a person places on a swastika will generate different emotions; throughout the Western world it is a sign of hatred, which people have learned to respond negatively to, and if a person is genetically predisposed to producing high levels of adrenaline they may be quick to anger when faced with the symbol. In contrast, throughout Asia the symbol has been associated with Hindu and Buddhist religions since far before the Nazis were ever conceived, and is understood to be associated with auspiciousness. Since every individual thing is the result of physical and cognitive events, which led up to that moment, everything which has ever occurred, or will ever occur, is interconnected in a predestined course of action which, with the collection of enough data, can be predicted all the way until the end of the universe, and possibly beyond. Everything that can possibly be imagined is a part of an equation, which defines the nature of and events of all things—an equation that is so immensely complex that it is likely that humans will never be able to even begin to understand it, but an equation none the less. There lies the primary challenge behind predictive analytics: understanding the interactions at work.

From a nearly limitless range of potential variables, many playing only a tiny role in predicting an outcome, one must identify those who have an influence in shaping those specific things being predicted. This can be a daunting process requiring vast volumes of information, and massive computing power capable of processing that much information under an automated system, which mines for significant relationships. If that is not difficult enough, the data required are not always necessarily readily available, particularly in predicting human behaviors, which change when a person is aware that he is being observed. Still, this process does not require a person to start from the beginning each time a new prediction is being made—once relationships have been identified and quality predictive models developed, they can be applied over and over again, being improved upon as new information becomes available. That means that this process is not one that requires starting-over, but is in a constant state of perpetual evolution, always becoming better than it was previously. Even in those things that do not show a significant relationship, this is not a failure since this, in itself, provides valuable information about things that do not have an influence on the outcome of an event. As noted by Thomas Edison, "I have not failed. I have just found 10,000 ways which won't work," until the right combination of filament, bulb, and vacuum were identified that would produce a consistent, long-lasting light bulb. More than simply providing information through the highly tedious process of elimination, this assists in identifying which variables can be managed without changing the outcome being addressed. The lack of significance can be just as useful as the existence of significance, when available to someone with a bit of imagination.

Another challenge, particularly related to geospatial intelligence and other abductive reasoning analytics, is finding ways to identify red flag observations. Currently, the most common method is through the use of mass-surveillance methods that analyze meta-data for trends in the usage of particular words. Prior to the outbreak of conflict or protest, there tends to be a spike over the Internet and on the phone for words associated with violence or criticism, for example. The real-time collection of global communications, however, has sparked controversy about the rights to privacy. While these have been mostly unheeded by the agencies, which utilize these surveillance methods, there has been a clear need for improved methods of screening for these red flags.

While predictive analytics are very useful in making predictions, as they are intended to, they do little for developing ways in which to respond. A prediction without a plan is akin to the myths of visiting an Oracle—you could learn of the future but do nothing to stop it. The conceptualization of an unchangeable fate assumes, however, that having knowledge is not a part of the determination—it separates the person from the external environment. While all events are outcomes of the events that led to that moment, acquiring knowledge through predictive analytics and altering the course of events is all a part of this. Predictive analytics only provide information about what will happen under a given set of circumstances. Then, once that has been accomplished, it becomes possible to alter the values of the predictive variables to predict what must be changed in order to generate a favorable outcome. A predictive model uses several variables, each of which can vary in value (hence the reason they are called variables, and not constants), so by calculating the outcome using several different combinations of values to determine which provides the best result, it becomes possible to develop strategies that accomplish these changes as their primary goals. This remains challenging, however, in that the analysis only shows how things must be changed, without providing any information about the way in which this can be accomplished. Analytics, as noted several times throughout this book, are merely measurements, and it takes human knowledge and creativity to utilize those measurements in ways that make functional change.

CHAPTER 11

Suggestions for Future Research

As with ongoing research into any type of analytical model, one of the most time-consuming but necessary operations to increasing the effectiveness and success of the analyses is the identification of relevant variables, and their proper incorporation into functional equations. Just like with descriptive analytics, this requires a great degree of somewhat tedious data mining, but with predictive analytics, the identification of these models and variables allow for the development of some very interesting tools.

First, it is effective when developing the infrastructure for ongoing analytical operations to build profiles for each variable and model. Each variable must have identification codes for the types of models and analyses for which they are a significant influence, just as it is in the nature of each model to identify the variables that are incorporated into it. Additionally, however, the profile of each variable must have a list of methods by which the value of that variable can be altered, and the efficacy of each method. The decision of whether or not the amount of value generated by altering the value of that variable in a specific model is worth the resource consumption and risk incursion could, of course, be determined by other analyses. Since a single variable will influence several different elements of any highly dynamic environment, it can be extremely useful to also identify these networks of influence, so that when incorporating the decision to change the value of a specific variable into the analysis, that the change in that variable will automatically generate assessments of how other elements of the environment will be influenced, illustrated in a web or matrix.

Not only does making these connection between models develop an increasingly accurate representation of the complex and dynamic combat environment, allowing for the first time in history strategists to have a comprehensive look at the varied myriad of influences a single decision will make, but it also allows for the management of timelines. Since each independent variable in a model is a dependent variable in some other model, and the predictions made by each model will be just another variable in some other model, the automated integration of analyses allows strategists to study each potential timeline that would result from any decision they might make. In weighing the various possible actions one might choose, every decision that is made will cause a different series of events to occur, which is sometimes commonly known as the butterfly effect. With the proper data and models available, each potential action and strategy can be evaluated in detail, and the long-term outcome of its implementation determined in advance. Conveniently, since a single variable is frequently used in multiple predictive models, this simplifies some of the integration.

As with all things in this book, this idea is not without supporting evidence, despite the almost unbelievable potential that it holds. In 1948, pacifist mathematician Lewis Fry Richardson developed an equation that modeled the progression of conflicts that remained consistent across all conflicts between 1809 and 1949, from minor skirmishes and rebellions, to WWII. More recently, Dr. Neil Johnson of University of Miami revisited this work and applied it to modern insurgency to test whether it still applies, and the model succeeded once again. This simple model, $T_n = T_1 n^{-b}$, calculates the amount of time that will pass before a specific cumulative number of attacks will be reached. In other words, the 50th attack (T_{50}) in a conflict is a function of the first attack (T_1) after a specified number of days, which increases logarithmically at the rate $-b$. While the exact determinants of $-b$ have not yet been identified, it is still possible to alter it as a relative value rather than an absolute value. There are a wealth of strategies that are used intentionally to speed-up/expedite, as well as slow-down/delay movements and attacks made by the opposition. As a simple example of timeline analysis, then, utilizing a model that predicts the impact of a specific strategy on the opposition can then be incorporated into a model estimating the amount of impact the result will have on the progression of a conflict, causing changes that create favorable conditions in specific variables associated with other strategic or predictive analyses. Should any military give analytics the attention they deserve and allocate a sufficient volume of resources to their development, they would have a tremendous advantage in being able to not only see the impacts on a secondary level, tertiary level, and well beyond, but even to actively manage the impact to facilitate desired results through careful management of the combat environment.

Predictive analytics, by themselves, can be immensely valuable—capable of predicting when and where a conflict will happen, where resources are being held, what the outcome of a battle will be and the number of resources that will be consumed, and identifying the variables that influence those outcomes so they can be managed to create a more favorable environment. Their true potential, though, lies not in individual analyses, but in connecting the analyses into an integrated series of dynamic cause and effect, which formulates a complex equation of the future.

PART III

Operational Analytics

Even being able to predict that an event will occur does little if nothing is done to properly manage a response. The success of any goal, the very survival of your organization, relies heavily on the ability to accurately assess the situation, develop a response, and execute it in a manner that is not only successful, but also efficient. Operations that are inconsistent or unreliable, functions that fail to efficiently perform, risks that are misidentified or otherwise overlooked, and incompetent allocation of resources can each be a harbinger of total collapse if left unchecked. By contrast, utilizing the proper analytics and responding appropriately to the data will generate a clear strategic map upon which the best path, and all pitfalls, will reveal themselves. The use of these analytics will identify specific strategic opportunities and threats, and, as has been shown countless times in the private sector, can be utilized as a strategic advantage on its own, in what has been recently dubbed "big data."

Being able to perform these analytics starts with a preliminary process called decomposition, which is the method by which all organizations and functions can be broken-down into a series of individual components. The resource-based view of the firm states that all components are a collection of inputs, processes, and outputs. The outputs from one component become the inputs of the next, in a never-ending cycle; and at every step the processes that are performed transform inputs into outputs which add value. The path by which resources take from their origin to the end-user with value added to them at each step is known as the value stream. The value stream includes not just the processes within a single organization, but also the processes of every organization in the supply chain, which is the basis of the SIPOC analysis. SIPOC stands for Supplier, Input, Process, Output, Customer; both supplier and customer can refer to components either inside or outside the same organization, and inputs and outputs can be tangible items or service offerings. SIPOC is a very broad overview of each components' place in the value chain used as a summary. More detailed analyses, called value-stream mapping, utilize specific calculation of resource consumption, value creation, and competing alternatives to that process, to develop strategy. Accountants will know this process as the same used in transfer pricing and in deciding whether to outsource a given operation.

As a portion of the deconstruction process, it is critical to understand the limitation of each component, and the limitations of the interactions between these components. There are primary and supporting operations—primary operations are those that are directly involved in the generation of a final output, while supporting operations are those that are necessary to facilitate the primary operations. Both create value, and both must have resources made available to them for goals to be achieved. The ability to turn inputs into outputs, though, is limited—an infinite amount of value cannot be created using limited input and limited time. The maximum amount of outputs each component can achieve is called their production possibilities frontier. This is the choice that each component must make in the utilization of their resources. They can produce a maximum number of one thing, or a maximum amount of another thing, or some lesser combination of both, but the total outputs will always be limited. These limitations are immensely important in determining the maximum capacity or potential over a given period of time.

Operational analytics do far more than simply identify limitations, though, as they are quite capable of expanding those limitations. The manner in which any organization is capable of utilizing their resources through a series of strategic processes, quickly, efficiently, without error, and without waste is fundamental to the ability of that organization to increase what they can accomplish, the speed with which it is accomplished, and the amount of risk to which they are exposed.

More than other types of analytics, operational analytics rely on a variety of non-statistical analyses. Many are still quantitative models that only incorporate statistical elements discussed throughout parts I and II of this book, while others are quantitative without any statistical elements, and still others are only graphical representations intended for brainstorming, identification, or process modeling. They all work in conjunction with each other, one supplementing the next, to provide critical information about the manner in which operations should be managed. Part I explained how to accurately assess a situation, part II explained how to predict what will come of a given situation, whereas part III defines how to respond to a situation with surgical precision, immense speed, and without error.

CHAPTER 12
==========

Quality Assessments
===================

Quality refers simply to a state in which operations or things exhibits the traits and functionality that they are supposed to, yet this concept that so many take for granted as an implied contract that should be minimally expected of the offer has been branded in many ways. The US Army states that it is their implementation of Six Sigma quality management, originally developed by Motorola, achieved a savings of $110 million during its third year of operations in 2005. Six Sigma, however, does not present anything that is particularly new or unique compared to its predecessor, Total Quality Management (TQM), which was developed by the US Navy in 1984. Even TQM is little more than a variation on the quality management analyses and tools developed by the American Society for Quality in 1946 as a way to bring quality management techniques by the US military during WWII to the private sector. Prior to WWII, even the origins of the modern bullet in 1846 has its roots in the perceived need to mass-produce a standardized, and consistently high-quality product, where prior to that it was common for individual manufacturers or even soldiers to fashion their own balls for use in muskets and pistols that had no standardized bore diameter. In fact, as far back in history as can be found it is common to find craftsmen demonstrating their skill and the quality of their work, particularly for those who could afford high-quality goods, which typically included royalty and knights, in their various incarnations. Granted, the modern methods of analysis and management have their roots in WWII, but it is clear that quality management has an immensely long history that is intertwined with military operations, which frequently required the mass production of a product upon which people could trust their lives.

It is also quite clear that despite the many ways in which quality has been branded and the emphasis supposedly placed on it in the military that quality is quite lacking in operations across a wide spectrum of operations. On February 14, 2014, reports of a bomb that was dropped in the wrong location as US military mistook their own people for the enemy, which would have killed several of them had the quality of the bomber's aim been any higher. Scandals reported as soldiers during OIF and OEF with clear mental disorders would knowingly attack innocent civilians, terrorizing them before killing them. Quality failings at Minot Air Force Base, North Dakota, during the spring of 2013 showed that not even nuclear weapons were exempt from the need for implementation in better quality management. The primary operations, ranging from launch officers and missileers of nuclear missiles, to the security forces which protect them, failed inspection miserably, while the overall inspection passed entirely because of the quality of the support staff——cooks, facilities management, and training programs. Unless one can reliably predict what the functionality and traits of operations and equipment will be, then the outcomes based on these things will be unreliable and a massive risk to the very survival of one's organization is born as a cancer from within its ranks.

At their core, all the approaches to quality management utilize something called statistical process control. Through the decomposition of an organization or process into a series of functions, each function in the value chain can be assessed for the rate of failure in terms of defective parts per million opportunities (DMPO). In other words, by sampling the output of each function before it reaches the next function, statistical analysis can be used to measure the number of defects per million units of output. The name Six Sigma refers to the number of standard deviations in a normal distribution utilized to calculate the confidence intervals that indicate the maximum rate of DMPO one can expect out of each process.

Should any function exceed that threshold, changes are made to bring the rate back down to an acceptable level. Some assessments then take an additional step and incorporate an adjustment in the number of standard deviations to account for long-term variations in quality levels. Six Sigma, for example, incorporates a 1.5 standard deviation adjustment shown in figure 12.1, which generates values shown in table 12.1.

LSL and USL refer to the lower and upper specification limits, while the blue and red distributions illustrate the 1.5 standard deviation of the mean.

The process capability index functions as an indicator for the likelihood that the 1.5 standard deviation of the mean will be exceeded. Process capability is a term that refers a processes ability to produce

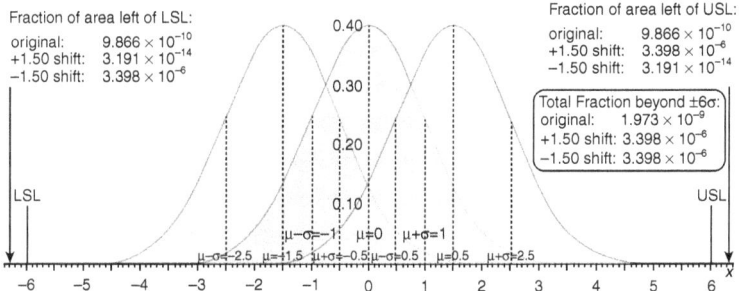

Figure 12.1 Six sigma distribution

Table 12.1 Six sigma statistical analysis

Sigma Level	1.5σ Shift	DPMO	Percent Defective (%)	Percentage Yield (%)	Short-term Process Capability Index	Long-term Process Capability Index
1	−0.5	691,462	69	31	0.33	−0.17
2	0.5	308,538	31	69	0.67	0.17
3	1.5	66,807	6.7	93.3	1.00	0.5
4	2.5	6,210	0.62	99.38	1.33	0.83
5	3.5	233	0.023	99.977	1.67	1.17
6	4.5	3.4	0.00034	99.99966	2.00	1.5
7	5.5	0.019	0.0000019	99.9999981	2.33	1.83

within specification limitations, and a higher number indicates higher consistency—lower variation.

Despite the emphasis placed on the statistical analyses associated with quality management, this composes only one element in the entire quality analysis process. Six Sigma calls this the "measure" phase, in the DMAIC phases: Define, Measure, Analyze, Improve, Control. By this point the individual functions and desired end result have been defined, and the statistical process briefly outlined above composes the "measure" portion. The "analyze" portion actually includes the greatest variety of analysis tools. The first of these analysis tools was developed by Henry Ford, and utilized setting physical specification limits. Ford believed that when a given operation exceeded these physical specifications, that loss would be experienced by the organization through product variation so great that either functionality was lost, error rates increased, or standardization was violated. This was later shown to be something of a misnomer, as the Taguchi Loss Function demonstrates, loss begins to occur any time specifications are anything less than optimal, but continue to increase at an increasing rate as production deviates further from optimal. Rather than upper and lower specification "goal posts" defined by Ford, Taguchi identified a simple parabolic curve, which illustrated the relationship between increases in cost and even small deviations from specifications.

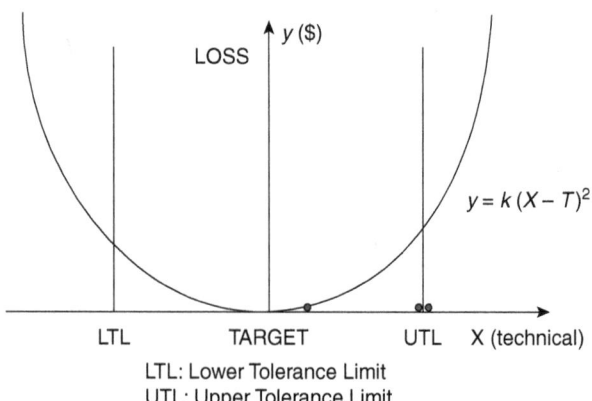

Figure 12.2 Quality goal posts

The critical-to-quality tree (CQT) is a modeling method that identifies specifically how to capture the end functionality or traits desired by the end user. The first step is to identify specifically who the target user of efforts is; who are you working toward benefitting, or who will be utilizing the output of your efforts? This can seem like an obvious thing that many people take for granted, but often there are subtle differences between target groups that have significantly differing needs. Ultimately, that is the point of identifying these groups: identifying their needs. Any production that does not work toward the needs of the end user is destined to fail in the functionality that is expected. The productive efforts of any organization must keep in mind what the output of their efforts are going to be used for, which is then used to identify the specific traits of the output the end user wants or expects. Remember that quality refers to the state in which something exhibits the functionality or traits expected, so if the output of your production does not exhibit those traits, then even if the defect rate is low, the quality will still be considered low because it is useless for the intended purpose. So, keeping in mind what the end user is trying to accomplish by utilizing your efforts, what traits must the output of your efforts exhibit? The end user will want several specific functions and traits, and will value each of them to varying degrees. Body armor will be considered low quality if it does not function properly in stopping a bullet, or if it is too heavy or bulky to move while wearing. The end user intends to protect themselves while remaining highly mobile in it, so it must exhibit traits of physical strength under impact, lightweight, and it must be small and flexible enough to allow the user to perform complex physical task.

The CQT is only the first step, though. Once that is developed, the information can be incorporated into something called a prioritization matrix, which is another statistical analysis tool, using weight prioritizations by trait. First, using a simple grid, the desired traits should be listed both along the rows and the columns, then using a 0–1 number system, identify the comparative priority of each compared to each other. So, in the grid, when the trait in a row meets the same trait in a column, there will be a null value, but when compared to the other traits the value provided will represent the importance of a given trait compared to the others. A 0 indicates that the end user would be willing to give up that trait in favor of the other, while a 1.0 indicates equal importance, and 2.0 means that the person is willing to give up the other trait to achieve the one in question, while the values range in-between. These values can be calculated by sampling end users regarding perceived importance, but calculating the rate in which each trait will be utilized, the severity of the failure of a given trait in terms of the total failure of the pursuits of the end user, and the current rate of failure as an indicator of urgency.

It is important, here, to note that perceived value of traits behave differently depending on the trait. Utilizing a Kano model, it is possible to more accurately understand the perception of different traits, and how the end users will respond to the degree to which each trait is implemented. Basic traits, for example, are those that do not provide additional satisfaction when it is fully implemented, but will generate negative satisfaction when it is lacking. In rifles, the consistency with which it correctly fires is considered a basic trait: when it does not misfire at all, it does not create additional satisfaction because that is simply what is expected, but when it misfires regularly it causes dissatisfaction. Performance needs are those that are considered key metrics in the production, and generate a linear relationship between implementation and satisfaction—the more, the better, at relatively equal rate. In rifles, such things as maneuverability, accuracy, firepower, and so forth are called considered performance traits. Finally, delighting traits are those that the end user does not generally consider on their own, and is considered a bonus feature that creates a lot of excitement. These traits increase satisfaction extremely fast the more they are implemented, but they have the caveat that the novelty of these traits will tend to wear-off over time and they will more resemble a performance trait in the long-run. In modern rifles, something like computer-guided aiming would be a delighting trait, but if a military force implemented these across all their operations then, over the years, people would simply come to expect them and become dissatisfied when they were not included.

It is also important to note that not all end users are equal, as all parties influenced by the operations in question will not have equal interest or equal influence. Each interested party is called a stakeholder, and not only must the traits important to the stakeholders be prioritized, so must the stakeholders themselves. A simple 4-square grid, which graphs the level of interest and level of influence called a stakeholder matrix, can be used to accomplish this. Each stakeholder is placed on the grid depending on their level of influence and level of interest in the operations. Those that have low interest and low influence will be apathetic to your operations, and should simply be monitored. Those with low interest but high influence will remain latent but should be kept satisfied and carefully watched for opportunities. Those with high interest but low influence are likely to compose the bulk of end users, and should be kept satisfied and informed to ensure they continue to defend operations. Those with high interest and high influence should be most carefully managed, and given highest priority. Since it is impossible to make everyone happy all the time, once the stakeholders have been graphed, they can be put into a comparative prioritization grid in the same manner as the traits.

Once the comparative prioritization values of the traits have been calculated, they are considered weights for a statistical correlative analysis (discussed in chapter 8) that calculates how strongly each option of action is successful in providing that trait. In other words, each option of productivity is analyzed for correlation with each trait, then the correlations are multiplied by the prioritization weights, to generate a priority index value. This is then also multiplied by the stakeholder prioritization multiplier. If a particular option offers more than one trait, then the index value for each trait it offers are added together. The total value of each option, then, is considered the priority value, which is then incorporated into decision-making processes, along with other quality tools, and other operational analytics described throughout part III of this book.

The traits associated with the chosen option must then be assessed and managed. The key to quality in these traits is that they will be consistently present and function properly each time. Variability, then, is considered a bad thing. The first thing that must be done to limit variability is to identify the types of failures to traits and functionality that can occur, accomplished utilizing a failure modes and effect analysis. For each trait or function, each potential point of failure is listed carefully, and organized for analysis (a simple spreadsheet works fine). Three aspects of that failure are assessed and quantified, and then those values are included in a formula to calculate yet another prioritization value known as the risk prioritization index. The first aspect is the potential impact of the failure, and what the results of that impact will be. If the impact will cause total failure of the pursuits of the end user, then this will have a much higher value than if the failure causes a decreased or temporary loss of functionality. If the impact creates harm to the end user in addition to simply ending their pursuits, then this value is even higher. The second aspect utilizes probability modeling (discussed in chapter 7) to estimate the likelihood of occurrence, which includes not just the many ways in which that failure might occur, but also the likelihood in which each way is encountered. A higher value, of course, indicates greater likelihood of occurrence. Finally, a detection value is estimated by estimating how easily the failure is to be detected, wherein a lower value indicates that the value is easy to detect and can be corrected before critical failure, and a higher value indicates that the failure is likely to surprise the end user. Any range of values can be used: 1–100, 0–1, and 1–10 are common value ranges; the importance is not so much on the values themselves, but that the values are consistently applied and are homogenous in distribution. When each value is determined, they are then multiplied together to generate the risk priority index, which provides important information about the degree to which each potential failure poses a risk to the quality of production. Risk management is discussed further in chapter 14.

Since variability has been the primary measure of quality failure throughout this chapter, it then makes sense that this variability should be carefully recorded and analyzed, which will be the final measure in the "analyze" portion of quality management. This can be accomplished using control charts, which utilize statistical deviations from the ideal specification, and utilize upper and lower control limits to identify any values that lie outside the ideal. The control chart measures the value of samples in a process and tracks them over time, and if values deviate outside the control limits at a rate greater than considered acceptable by the benchmark DPMO rate utilized in statistical quality measurements, then something is wrong with the process being measured, and it must be fixed. Wild or inconsistent deviations in the process are the result of inconsistencies within that process, and it must be refined. Consistent deviations that stay outside the specification limit or grow increasingly deviated, called random walks, are the result of the process having changed somewhat and must be reset.

Whether a quality assessment or process or evaluation or other type of event is good or bad, the ability to improve upon poor quality and understand good quality requires an assessment of what caused the quality to become good or bad. A common tool to accomplish improving quality is a cause-and-effect diagram, commonly called a fishbone diagram because of its shape when it is drawn. The head of the fish is the end effect, from which a single spine is drawn, then four ribs sticking out of the spine are labeled as policies, procedures, plant, and people—each of the potential sources of the current level of quality. Then out from the ribs are varying numbers of smaller bones, identifying each cause of the level of quality associated with the people, policies, and so on. Then out from those "cause" bones, are tiny ancillary bones identifying why that cause influenced the quality. In the end, the entire thing looks like a fish skeleton, but the function is to demonstrate how policies, and so on cause changes to quality, in order to identify the influences on quality within operations and improve upon those creating negative quality.

If the sources of quality problems cannot be identified, they cannot be improved upon. After all this analysis, it is unlikely that any issues remaining in quality will be left unidentified. Should that be the case, however, or if those improvements still prove to be insufficient, then benchmarking can be a useful method for limiting weaknesses within any process. Benchmarking is the process of identifying which organization performs a process more effectively than your own organization is capable, and then copying them. It is a pragmatic way to ensure that operations remain competitive, and that any operational sources of strategic weaknesses are at least as good as those that can perform it best (discussed more in chapter 13 on "Efficiency Analysis").

William Edward Deming is often credited with popularizing the modern global quality movement, having started promoting quality improvement methods in the late 1920s, and teaching it broadly in Japan starting in the 1950s, which likely contributed much to the current Six Sigma approach. He popularized the PDCA cycle which refers to the cycle of processes: Plan, Do, Check, Act. In the final step of quality management, after improving the quality problem, it is still necessary to control processes to ensure that quality stays high or improves. This is the point in which the cycle restarts, wherein planning begins again by defining the problems, end users, processes, and so forth. Deming demonstrated how quality is not a cost, but that organizations cannot afford to not pay for quality. Low quality results in defective work which never makes it to market and must either be scrapped or reworked. It results in work that fails after reaching the end user and results in lost confidence, recalls, lawsuits, or process failures. It results in reevaluation of processes and materials, and it results in a loss of economic resource availability or resource utilization efficiency thereby increasing costs of resource acquisition and utilization (e.g., a low quality process that harms people in down the value chain in operations will cause them to be less productive, increase the cost of compensation, and increase demand in the labor market for higher rates of compensation to perform that function). Deming is famous for his 14 principles and 7 deadly diseases of quality, but in the end, the entirety of this chapter is summarized in one simple calculation:

Value: Usefulness/Cost

Without quality, processes have less usefulness and, therefore, less value to anyone. The more variable quality of an operation is, the more it contributes to lower levels of average usefulness.

CHAPTER 13

Efficiency Analyses

The immense military success of Alexander the Great is largely attributed to the methods he used, which focused on improving speed and mobility. Alexander kept his army comparatively small, required regular training in sustained marches, and focused heavily on reducing the volume of servants and the size of the baggage train, greatly increasing the army's marching speed. This enabled him to frequently arrive before his opponents had anticipated. Alexander was also fond of a "hammer and anvil" combat strategy, which employed the use of elite forces specially trained to be swift to flank behind the opposition, thereby attacking from in front and from behind. He was known for making decisions quickly, and the speed of his conquest was without compare—moving from one to the next seemingly without delay. It was this speed that allowed him to initially gain control over a unified Greece, portions of which were in open rebellion soon after Alexander took the throne in Macedonia, as his opponents were stunned by the speed with which the Macedonian army conquered their opponents.

In a similar manner, it was speed that allowed Nazi Germany to be so successful in WWII. Their Blitzkrieg combat strategy ensured that their conquests were won extremely quickly by overwhelming their opponents, and avoided the problems of attrition associated with the trench warfare of WWI, of which Germany was all too familiar. Nazi Germany was also among the most successful to mobilize the power of industrial warfare, being one of the first to utilize modern methods of standardized mass production utilizing assembly lines and mechanized factories in order to both supply the war effort, and to conduct the war, itself. Later in the war, the United States came to pride itself on its ability to quickly and powerfully utilize similar methods by transforming automotive plants into factories of national defense. This success is attributable to the ability of a nation to use the resources available to it in an extremely efficient manner, which ensured amply supplies, the availability of large capital equipment, and a speed in both movement and strategy that were completely unprecedented in any war throughout history up until that point.

It is not enough to simply assume that operations are efficient, or as efficient as they can be, and to claim otherwise is just guessing as casual observation is highly subjective and flawed. The use of detailed analysis frequently identifies inefficiencies previously unknown, helps in improving future efficiency, and contributes to the planning of operations and missions to ensure they will be as fast and efficient as they possibly can be. There are many different types of efficiency, each contributing to improving operations, and each requiring their own type of special analysis.

When talking about efficiency, people tend to think exclusively about productive efficiency, which refers to how effectively an organization is utilizing their resources. The formal definition of productive efficiency is the degree of productivity wherein Marginal Cost = Average Total Cost (MC = ATC). The marginal cost of production is the cost of +1, the cost of one additional unit of productivity, and marginal cost increases with increased production as additional resources are consumed and become more expensive. Since any operation requires an initial expenditure even before operations can begin, that means the average total cost will start very high, as the average cost of producing 1 unit will include not just the production cost but also the entirety of the initial costs associated with getting started. The second marginal unit of production still has production costs associated with it, but now the initial start-up costs are distributed across two units of production, instead of just 1, bringing the average total costs down. ATC will continue to decrease as long at MC is lower, but as ATC decreases and MC increase, MC will eventually surpass ATC causing the average to increase, again. It is the point where MC = ATC that ATC will be lowest. It is also at this point that current assets have reached their productive capacity, and either additional resources must be consumed, thereby increasing ATC, or current assets must be replaced with assets that have higher productive capabilities.

This same basic premise also applies to human resources, as well, although in a slight variation. There are, of course, simple calculations that can be done to determine the average productivity per labor hour (Total Production/Number of Labor Hours), the average resource costs of labor per hour (Total Compensation/Number of Labor Hours), and the average net benefit per employee (Average Productivity – Average Resource Cost). You can also assess labor management by evaluating the difference in labor utilization and labor needs, by measuring production requirements during a period, dividing that by productivity per person, and assessing the difference to see whether operations were overstaffed or understaffed during a given period. Of course, these things can all be utilized in estimating the future needs of human resources, as well, but there is a key difference between human resources and other types of resources: variable productivity.

In any piece of equipment, the rate of production and total amount of production are both fixed—they are constant values, which are known in advance and must simply be accepted as a limitation of the resource. Human resources, however, have the ability to learn, adapt, innovate, and generally improve. For example, when using cross-functional teams, or cross-functional training, a person can better understand the needs and capabilities of those with whom they work, and can consider how to improve the users of their own efforts, or what the suppliers of their resources can do to improve the work process. People are subject to variations in productivity and creativity based on changes in motivational factors and the environment in which they work. A critical component of efficiency, then, is understand what makes people unique, and how best to utilize their efforts. Generally speaking, people are capable of making decision and utilizing their imagination. We are capable of asking "What if?" People are capable of creating systematic improvements, while other forms of resources simply act within the parameters of the system developed by people. As a general rule, any role that requires critical thinking or creativity can be done more effectively by people, while roles that are repetitive and function only as an automated response to stimuli are more efficient when automated by computers and machines. In some cases, the mobile nature of operations, or the physical logistics of location, make it difficult for automation to be implemented, but these limitations are diminishing with global remote functionality, such as with the operations of drones and, more recently, infantry robots.

There is a matter of productive efficiency that must be considered, then, in the balance between capital intensity vs. labor intensity in operations. For any given operation, the average productive efficiency of labor and capital must be compared, through calculations of factor productivity. The factors of production include land (the physical land and the things in it or on it), labor (workers and management), capital (physical things), and entrepreneurship (the willingness to accept risk to generate benefits). Any calculation of factor productivity includes a core ratio of outputs to inputs. For example, as already noted, labor productivity is the ratio of total production/total labor hours. Capital productivity would be calculated as total production/capital resource consumption. Total factor productivity is calculated as total outputs/total inputs, and is used as a benchmark comparison for the productivity of individual factors. The balance between labor and capital is met when replacing any amount of one with the other decreases total factor productivity. This must also include the degree of risk incurred, since high-risk situation, such as combat, carries the potential to consume that resource in a single moment (i.e., being destroyed), and people carry with them the necessary costs of long-term compensation if injured or killed, whereas machines are simply rebuilt.

Another form of efficiency, similar to productive efficiency, is called distributive efficiency. Distributive efficiency is achieved when resources are distributed in such a way that they create the maximum amount of benefit. This differs slightly from productive efficiency in that you are not just considering the volume of usage of these resources, but also the projects they are being allocated. Determine the potential value of each project (discussed in more detail in chapter 15, on "Asset Management"), and the resource needs of each project, so that resources are distributed to projects creating the most value, allowing the resources themselves to improve their own functional value. To say simply that a given mission or project has the highest value and that sending all available resources to that project creates the highest value is incorrect, though. Distributive efficiency takes into account certain productive laws. First, the law of diminishing marginal returns states that each marginal unit of consumption will generate less marginal benefit. In other words, when you send more resources than needed to complete a project, those resources may create some benefit per unit, increasing total benefits, but the benefits of the extra resources will be lower than the resources needed. A good way to look at this is with infantry: If you can accomplish the mission with a minimum 50 infantry, then each of those people is generating the maximum amount of value possible in that mission. If you send, instead, 100 infantry to complete the mission, then the additional 50 people will still create value, but that value will be less than the initial 50 simply because there will be less for them to do. Then, if you send 500 infantry, suddenly there are not enough resources to sustain them all and you are actually creating negative benefits.

Of course, in sending just a few extra, they're still contributing some positive value, so if you have the extra people, achieving distributive efficiency may mean that it is best to utilize them as extra rather than in any other potential use (e.g., in the infantry example, sending some supplemental people to the mission may be a better use than performing janitorial work at the forward-operating-base (FOB). The question to this can be answered by determining the marginal rate of substitution (MRS), which measures one option as a ratio of another option. If you have the choice between two projects to which you can allocate resources, MRS would require you to figure out how much you can accomplish by allocating varying volumes of resources to either of them (discussed further in chapter 15 on "Asset Management"). So, allocating 60 percent of resources to option A would leave 40 percent of resources to option B, and the amount of productivity generated by each, then, could be divided to calculate a ratio that describes how much of one would be needed to replace the other. Distributive efficiency is achieved, then, when these are equal.

A form of efficiency that can, in itself, directly contribute to overall missions strategies is called Pareto efficiency, which is achieved when resources cannot be allocated to one function without taking them away from another function. In other words, all resources are being utilized, and it is only possible to improve Pareto efficiency if one can be benefitted without creating any detriment to another. Its counterpart, Kaldor-Hicks efficiency, functions in a similar manner, except that it allows for detriment to one function when the increased productivity from the operation being improved can at least equally compensate the function experiencing a detriment from the shift. This is utilized in strategy to determine relative comparative advantages between various allies and opponents, and generate a competitive advantage.

An absolute advantage exists when one side is able to do everything better than another, but this would require one to completely overwhelm the opposition, which is not really a strategy at all nor is it sustainable, as proven during OIF and OEF. Comparative advantage, however, looks at what each side is capable of achieving by consuming the fewest resources by considering Pareto efficiency. Consider tables 13.1 and 13.2.

Table 13.1 Productivity potential without cooperation

W/out Cooperation	Ally A	Ally B	Total
Function 1	50	33	83
Function 2	25	99	124
Total	75	132	207

Table 13.2 Productivity potential with cooperation

W/ Cooperation	Ally A	Ally B	Total
Function 1	100	0	100
Function 2	0	165	165
Total	100	165	265

By cooperating with allies to focus on the operations in which each is most effective, much more can be accomplished far more efficiently and far more quickly. Functionally, this can be measured by calculating transfer pricing. By deconstructing the organization and measuring how many resources are consumed by the various functions as each operates toward a single purpose, those functions that consume more resources, or have higher costs, than another organization that can perform a comparable function utilizing fewer resources should be outsourced. By assessing comparative advantages against the opposition, it is possible to determine what function each side of a conflict is more competitive in. One is now able to more carefully understand the strengths and weaknesses of the opposition, and one's own strengths and weaknesses, and develop strategies that emphasize the differences between these. Once the absolute advantages have been identified, they can be incorporated into other functions and utilized in primary operations as a strategic advantage.

When Kaldor-Hick efficiency is combined with distributive efficiency, the focus becomes utilizing all resources in a manner that is most beneficial, which is a state known as allocative efficiency. Allocative efficiency is different than Pareto efficiency in that allocative efficiency considers optimal use of resources, rather than maximum use. Then, in combination with productive efficiency, one is ensuring that all resources are achieving their maximum potential. Once these are calculated, it is possible to determine what you are capable of accomplishing, even if it is yet to be accomplished. With all these different considerations to weigh, each utilized to narrow-down the best course of action and improve functionality, it can become easy to narrow the focus of analysis too much and lose sight of long-term strategy when dealing with short-term operations. Dynamic efficiency looks at the needs of short-run and long-run planning, and utilizes the same basic assessments utilized in Kaldor-Hicks efficiency so that any loss of long-term needs is compensated for in short-term gains. Then, when an organization is capable of achieving maximum efficiency, but fails to as a result of less than optimal operating practices, an *x*-inefficiency is generated. The *x*-inefficiency is essentially inefficiencies that result either from human error, or a situation in which it is more beneficial to be inefficient.

The final measurement of efficiency is to ensure that any gains in efficiency are efficient. Increasing efficiency will improve speed, reduce resource consumption thereby increasing the availability of resources, increases the effectiveness of operations, and reduces risk. Yet, it is possible to reach a point that so much time and so many resources are being consumed with efficiency management, that it becomes inefficient. It is important that efficiency operations are subject to the same analyses used to assess everything else. Efficiency operations are optimized once the additional cost of expanding efficiency operations is equal to the gains experienced. Any less, and it is possible for operations to be more efficient, but any more and efficiency operations stop improving operations and begin hindering them.

One of the primary benefits of efficiency is increased speed of operations. As fewer resources are required, the effectiveness with which they are used improves, and the focus is shifted to accomplishing goals as quickly as possible in order to minimize resource utilization, this will naturally speed-up the rate with which operations occur. This also has the added benefit of decreasing the amount of risk to which one is exposed. As missions are completed more quickly, there will be less time in which they are exposed to risk, and the reduced volume of resources necessary to accomplish a mission means that fewer resources will be put into risky situations in the first place. By reducing risk, the amount of excess resources needed to account for that risk also decreases, and the two benefits supplement each other well. In order to maximize these benefits, there are some analytical tools available, relying entirely on the analysis of the operating cycle. For our purposes, the operating cycle begins when a decision has been made to pursue a project, then resources are acquired, those resources are utilized to create project-necessary assets (i.e.: inventory) available to begin primary operations, primary operations are executed, gains from those operations are realized, then those gains are utilized as an asset in the pursuit of the next project, at which point the cycle begins again. The operating cycle measures the duration of time, in days or years, which is required in order to go through the entire cycle once, from beginning to end.

In 1957, an analysis tool known as PERT (Program Evaluation and Review Technique) of the operating cycle was developed for the US Navy Special Projects Office as a way of managing the development and production of the Polaris nuclear submarine project. Such a complex project requiring many extremely complex tasks was difficult to effectively grasp and manage in terms of project management; so the tool was developed to break-down the process into several major steps. Each major task identified was utilized as a unique project in itself, and statistical analysis (such as those described in parts I and II of this book) were utilized to estimate the amount of time necessary to complete each task. The best estimate was given, as was the optimistic estimate, and the pessimistic estimate (which were the upper and lower limits on a defined confidence interval). Once that is done, then it became possible to generate certain key valuations, such as the critical path—the longest potential time needed to complete the project, used as a conservative estimate of project time requirements. To account for production time variations, the expected time to complete each task was calculated as a weighted mean of the pessimistic, optimistic, and best estimates of completion date: (Optimistic+Pessimistic+[4×Best])/6. Then this, and an understanding of one's ability to perform several tasks simultaneously (called fast-tracking), and the earliest time in which the next task can begin after the previous is completed, or shortly before (called lag time), an accurate analysis of the total time requirements of the project could be produced. With a specified deadline, it also became possible to estimate the amount of slack available (any additional amount of time or resources in excess of that available), and the lead time of each task as continuously assessed throughout the operating cycle (the amount of time required for each task to be completed if the next is to be completed on schedule.

Once the PERT analysis is completed, then the data are utilized to develop a Gantt chart. The Gantt chart is a simple spreadsheet-based bar chart that illustrates the relationship between each task in a project, and dates. The bar itself is broken into multiple components illustrating the estimated time, minimum time, and maximum time of completion. This is a simple way to visually manage the progress of a project throughout the operating cycle. Over time, and over several projects, by utilizing efficiency analytics, it is possible to track changes in the efficiency of operations so that specific operations can be identified either as strengths, or functions that require additional efficiency emphasis. This also helps to identify systematic bottlenecks, wherein operations are being slowed by a specific function which may require changes to the assets available, or even increasing the number of units performing the same function.

Another major benefit to increasing efficiency is that resource consumption is reduced. As projects are completed more quickly, fewer time-dependent resources are consumed, requiring fewer resources to be transported and kept on inventory. Another way to look at it is that given resource and logistical limitations, or relative resource cost constraints, the lower consumption of resources increases the amount of resources available for additional projects or for allocating to improving returns on specific functions. The lower consumption of resources also means that the resources that are utilized are accomplishing more per unit, indicating increased effectiveness of operations. Just as time benefits can be managed by understanding the operating cycle, resource benefits can be managed by utilizing materials requirement planning (MRP). MRP refers to any system in which the resources necessary to maintain operations are predicted, sourced, and controlled. Modern innovations in MRP have been seen most greatly in the private sector such as in large retailers, such as Wal-Mart, whose sales and inventory tracking systems are all automated and nearly instantaneous. The instant that a purchase is made, the exact items being purchased are communicated to suppliers as an order for their next shipment, while sales data are stored in the inventory system and analyzed by computer to identify fluctuations in purchases based on a variety of factors such as geographic location, demographics, chronological trends, and seasonal fluctuations.

The volume, frequency, and types of orders being made are subject to analyses related to economic order quantity (EOQ), which accounts for the costs of several variables and maximizing the total benefits derived from the available resources by decreasing inefficiencies related to resource management. These costs include the costs of the resources themselves, the cost associated with making the order and shipping the resources each time, and the costs of storing resources once they are shipped. So, the total cost is equal to:

TC = (Price × Quantity) + Ordering and Shipping + (Quantity × Storage)

The goal, then, is to reduce the number of shipments ordered as much as possible, without increasing the amount of resources in storage so much that they become more expensive than simply paying for extra shipments. The economic order quantity, or the optimized volume of orders, can be calculated as:

EOQ = √(2 × Cost per Order × Annual Order Quantity)/Storage Cost per Unit

This is also supplemented by a management technique of inventory and production known as Just-in-Time (JIT). JIT utilizes indicators, called Kanban, in each step of the operating cycle to indicate to the previous step that more in-process inventory is needed. In other words, with a production process that has three steps, once the second step is nearing completion a signal would be sent to the department performing the first step to produce another unit. The amount of time required to perform a given function, or to order and ship additional raw inventories, is taken into account so that storage costs are minimized. This means that inventory and in-process inventory arrives just in time for it to be needed, and no sooner, so that the actual amount of inventory in storage is kept to nearly zero at all times. This also reduces the total length of the operating cycle, because the amount of time that materials are kept in storage before their final use is dramatically reduced, thereby further increasing speed, reducing risk, and improving total resource consumption.

Ratios available to calculate resource utilization and track changes in resource efficiency are called "turnover." Inventory turnover, for example, is calculated first by determining the average value of inventories: (Current Period Inventory + Previous Period Inventory)/2. Then, using average inventories, inventory turnover is calculated as: Cost of Operations/Average Inventory. This effectively calculates the number of times in a period that inventories are fully consumed and replenished, known as turning-over. Slightly different, asset turnover calculates the productivity generated by assets, using the ratio: Production Value/Total Assets. The more production that is created as a ratio of assets, the more effectively those assets are being utilized.

In the 1990s, the term "Lean" began gaining immense popularity as a method of increasing organizational efficiency in all types of operations. The term came to summarize the methods utilized in the Toyota Production System (TPS), as Toyota was seen, at the time, to be a benchmark in operational efficiency. The roots of lean, though, stem from the early twentieth century and the original assembly line methods of mass production. When asked about his inspiration for TPS, creator Taiichi Ohno stated that he had learned it from Henry Ford's book, *Today and Tomorrow*. The emphasis of lean operations is to increase process velocity, quantify and eliminate waste, remove complexity, identify and minimize activities, which do not add direct value to the end result, and use efficiency tools to analyze production flows to identify delays. There are eight forms of waste that this seek to remove from operations incorporated into the acronym DOWNTIME: Defects, Overproduction, Waiting, Non-Utilized Talent, Transportation, Inventory-in-Process, Motion (extraneous), and Extra Processing.

In order to eliminate waste and remain competitive, lean requires three broad steps. The first is systems reengineering. The entire operations process must be analyzed in detail, mapped-out, quantitatively measured, and redesigned. This falls into the second step, which is to recognize that there is always room for improvement. With each function within a system, the focus must be placed on what can be improved, changed, or eliminated. Can an entire function be bypassed or replaced? If not, what can be done to make that function better? The final step has become known as continuous improvement, which is indicative of a learning organization. This means that everyone within an organization, at all levels, are collecting information and seeking ways to improve the overall operating process at all times. By constantly accepting feedback and identifying inefficiencies, an organization maintains a perpetual state of evolution rather than ever requiring major overhauls that are indicative of a failure to properly respond to internal and external influences.

Through this continuous improvement, the operating cycle will gradually decrease in duration and resource requirements, while simultaneously improving in quality and effectiveness. In addition, the process of improvement will gradually improve, as well. There will be time cycle variations with each progressive succession of system improvement, which will not only improve the operations cycle but also increase the rate at which the operating cycle improves. Any force which can operate more efficiently than its opposition will find that the speed and effectiveness with which it operates are a huge competitive advantage, just as past militaries from Alexander the Great to WWII have achieved the seemingly impossible by becoming unrivaled in their operational efficiency.

CHAPTER 14

Risk Management

Every potential decision we make is fraught with a variety of risks, and the decision to not act at all is frequently the most dangerous. In complex and dynamic environments there are huge volumes of decisions that must be made at each moment, with the number of potential responses, or lack thereof, limited only by the imagination of those facing the choice. The very success of an operation depends largely on one's ability to properly assess these risks, yet without an understanding of the methods available to accurately identify and measure those risks inherent in each potential response, a person is just guessing at which will have the best outcome, or whether a response will even be successful. In the best of circumstances, such haphazard methods of managing operations will result in less than optimal decisions that create inefficiencies, but when you include important resources, the success of an operation, or even the lives of people in the decision you make, the accuracy with which you manage risk could literally be a matter of life and death. By applying statistical measures already discussed throughout many earlier chapters to a given operating environment and the resources that will be utilized within that environment, though, it becomes possible to turn these risks into a potent strategic advantage by managing the exposure of each side of a conflict to these risks, and by being able to improve one's response and assessments of strength after encountering a risky situation.

Although there are several different models of the risk management cycle, they are all variations on the same five basic steps. The first step is considered part of the assessment process; it specifically identifies each of the threats to which one will be exposed, including the development of a profile of the nature of that threat. The second step in the assessment process is to carefully measure each of the threats, which includes measuring not just the probability of each occurring, but also doing detailed analyses of the implications of each threat, thereby allowing for the development of a prioritization schedule based on comparisons of these measures. The remaining three steps to risk management are considered a part of the management process, starting by making a decision based on the measurements that were performed, which includes not only choosing the best in any given decision, but also the responses that will be taken to each of the risks inherent in that choice. Next, controls are executed to optimize the operations that stem from the decision that was made, responding appropriately to each threat. Finally, evaluations are made of the manner in which these risks were managed, and improvements made to the risk management process for the next decision that will have to be made. This triggers the cycle to repeat, so that threats are identified in the next decision utilizing any improvements that were made as a result of lessons learned in the previous cycle.

When reading through the literature on risk, there is some disagreement about the nature of risk. Many organizations identify risk as the degree of variation or volatility to which one is exposed, as defined as

$\beta = \text{Cov}[a, b]/\sigma^2_b$

This is called Beta, which measures the amount of variable response that a specific object has to changes in the same variable in the environment around it by using statistical covariance between the specific object and the object's environment. In other words, if violence in Afghanistan changes by 1 percent, and violence in Kandahar changes by 2 percent, then Kandahar would be considered more volatile. This view of risk lacks validity, though, since even severe and frequent changes in the comparative strategic advantages between opposing organizations, or a large degree of volatility in the violence of a region, are still preferable to maintaining a constant disadvantage in a region under constant threat. A 2 percent change in the violence of Kandahar in response to a 1 percent change in violence in Afghanistan may actually mean Kandahar is decreasing in violence more quickly, rather than increasing in violence. Such fluctuations can even create a strategic advantage, if one can accurately predict the timing of these changes and respond more efficiently than their competition. Volatility, then, is not a form of risk, in itself, but it does pose the potential to create a degree of uncertainty as compared to a more predictable environment. Uncertainty is another definition used to describe risk in some literature, such as the ISO Guide 73, which defines risk as, "the effect of uncertainty on objectives." While uncertainty is one type of risk, potentially obscuring the threats one might encounter, a person can be quite certain of the threats they will face without reducing the amount of risk it poses even slightly. So, uncertainty, as well, is insufficient for our purposes. Instead, the definition of risk that will be utilized here is a more precise one: the value or volume of losses resulting from a threat as measured by the probability of encountering that threat and the degree of exposure.

With a functionally useful view of what risk is, it becomes possible to begin the assessment process. When faced with a decision, begin by developing as many response options as possible. Then for each, develop a model of the context. According to the Institute of Risk Management (IRM), there are two broad sources of risk: internal and external. Internal sources of risk are those that are caused by one's own management and operations. These can be identified by assessing the various functions that will be performed, the people and resources who will be performing them, the policies and people who will be managing these functions, and the goals that they are attempting to accomplish. These risks might be caused by human error, insufficient safety considerations in the design of equipment, and other things resulting from one's own organization. External sources of risk are those that are caused by the environment in which you are operating. These can be identified by considering where you will be operating, when you will be there, what that area consists of, and what else will be happening there. These risks might be caused by natural disasters or inclement weather, or by proximity to opposition forces, among many other possible sources. Once these sources of risk have been identified, it must be determined for each the time-horizon of the influence they will have. The IRM identifies three time-horizons: Short-term risks are those that will impact operations, medium-term risks are those that will impact tactics, and long-term risks are those that will impact strategies. So, in other words, a risk which will not stop the success of any single battle but will cause a persistent disadvantage or otherwise derail your current strategy is considered long-term. For example, a shift in alliances, such as that experienced during OEF when the Afghan government decided they would no longer cooperate with foreign forces nor approve of their continued occupation in Afghanistan created a long-term change to the strategies of those foreign forces. Medium-term risks during OEF and OIF necessitated changes in tactical responses, since any systematic response to an event allowed opposition forces to anticipate the actions that would be taken, such as a consistent method for approaching villages allowing for the development of a predictably effective ambush, or the placement and timing of multistage roadside bombs resulting from observed stops and movements in response to the detonation of the first. Short-term risks would be those that impact operations, such as the risk of getting shot in the midst of combat, which would certainly hinder the ability of the injured individual to continue mission operations.

Together, the source and time-horizon provide critical information about the nature of each type of risk, but does not describe the types, themselves. Many types of risks are unique to specific industries, but, although military operations face risks that tend to be more severe than in most industries, the types of risks faced broadly translate to most.

- Operating Risk: The risks that result from your primary operations. Even within the military, these can vary quite widely. Truck drivers have a higher than average risk of being involved in a traffic accident, which would be a part of the primary operations of a transportation unit, while an infantry unit has, as a part of their operating risk, the risk of being shot.

- Attrition Risk: Attrition occurs when resources or capabilities are gradually diminished through extended operations. In a war of attrition, each side may very likely experience shortages. Another form of attrition more common in civilian industries is called interest rate risk, or inflationary risk, wherein the value of one's capital resources slowly diminishes in its ability to fund operations as a result of inflation.
- Liquidity Risk: Liquidity is the ability to turn assets into usable resources. For example, cash is extremely liquid, but if you invest in the construction of a building, it would take a very long time to then resell should you desperately need to purchase ammunition, instead. A lot of military assets are tied-up in construction or contracts, such as large industrial construction of ships, which can make reallocation or budgeting difficult in emergency situations.
- Hazard Risk: General hazards are those experienced by most people. The risks associated with natural disasters or inclement weather, for example. Hot weather and cold weather injuries are quite common in the military, and are given a high level of attention due to their frequency and potential impact on mission readiness.
- Strategic Risk: These are the risks associated with changing strategic and competitive environments. The threat of new entrants, the treat of disruptive technology becoming available to the competition, the threat of political changes, or the threat of shifts in supply chains. There are many possibilities, but the focus is on long-term dynamics that alter one's competitive advantage.
- Uncertainty Risk: Uncertainty is a word that refers to having imperfect information. Imperfect information brings with it a degree of probability of miscalculation in the risk assessment process; either a risk was not properly identified, or it was not properly measured. Complex and dynamic environments often come with a high degree of uncertainty, which is partially why carefully understanding the information to which you do have access is so important.

To help simplify this process, risks that are commonly encountered with little variation, or variation that is highly predictable using one or two correlative variables, can be categorized for easy reference and reuse.

Having identified the risks associated with each potential choice of response to a given decision does nothing for one's ability to make a proper decision. Simply because one option has more types of risks does not mean it is any better or worse than an option with only one

Severity \ Probability		Frequent A	Likely B	Occasional C	Seldom D	Unlikely E
Catastrophic	I	E	E	H	H	M
Critical	II	E	H	H	M	L
Moderate	III	H	M	M	L	L
Negligible	IV	M	L	L	L	L

Figure 14.1 Composite risk management

type of risk, because questions remain about the amount of threat posed by each risk. In order to compare the amount of risk associated with each option to a decision, first you must measure several variables associated with each risk. The current military model for doing this is called the risk assessment matrix, shown in figure 14.1.

This matrix broadly assesses the severity and probability of each type of risk in order to define the risk level associated with them. Clearly, this is not done quantitatively, but by casual observation and the best guesses of the person assessing risk. The levels of each are as follows:

Probability

- Frequent: Continuously encountered
- Likely: Regularly encountered
- Occasional: Sporadically encountered
- Seldom: Unlikely but not impossible
- Unlikely: Assume it will not occur

Severity

- Catastrophic: Death/disability, or systems and property loss if encountered
- Critical: Major injury, system damage, or impairment to mission if encountered
- Moderate: Minor injury, impairment to mission if encountered
- Negligible: First aid, no impairment to mission if encountered

Risk Level

- Extremely High (E): Loss of ability to accomplish mission
- High (H): Significant degradation of capabilities
- Moderate (M): Minor injury, system damage, or property damage
- Low (L): Little or no impact

Clearly, this method of risk measurement is prone to an extremely high degree to subjectivity and error, and does not include, at all, the resource costs associated with each potential option. It does little to accurately measure the true nature of risk, much less provide information about what to expect or mission readiness after such an encounter, and the amount of uncertainty risk to which you are exposing yourself by using this method is "extremely high." It does have the benefit of being simple to use, however. It can be performed in a moment's notice, with little thought, and without a prerequisite understanding of analytics. Until recently, such ease of use was a critical factor for any tactical or strategic assessment tools intended for use in a field situation. More recently, however, technology has become available that allows people to perform extremely advanced and complex analytics quickly, using data accessible in huge global databases and information centers, and still without having any prerequisite understanding of analytics. To quote the Apple marketing campaign for the iPhone, "There's an app for that," or at least there can be. The data and analytics that can change the combat theater are easily adaptable, and frequently no different from those used in standard business intelligence software commercially available from IBM, SAS, and other companies. Our ability to measure risk, then, has completely changed in its nature, and can now provide extremely precise information and comparisons.

When managing risk, it is necessary to understand the relationship between risk and reward, on which many people get confused. More risk does not inherently translate into more reward. It is quite the opposite—in order to make it worth accepting the additional risk, there must be the promise of additional reward. Given two choices that have equivalent potential benefits, and will cost the same to pursue, if one has a lower amount of risk associated with it, then that is the best option. The concept of "no guts, no glory" is a misnomer, as any risk can be easily calculated as a resource cost, using methods that will be discussed throughout much of the remainder of this chapter. The basis is that by calculating the volume of resources that are being exposed to a particular risk, and calculating the probability of that risk occurring, an estimation of the amount of resources consumed by encountering that risk is derived.

One of the more common calculations of the cost of risk is one originally developed for financial investing before attracting broader industries and their related risk management applications: Value at Risk (VaR). Recall from chapter 7 ("Probability Modeling"), standard deviations in a data distribution can be used to estimate the degree of certainty you have that actual values will fall within a specified range. Of course, these distributions are two-tailed, but in risk management only the bottom tail, the portion of the distribution representing the degree of certainty of loss, is relevant. By applying this to resource cost-based risk assessments, this can be used to estimate within a given confidence level the most that will be lost. Say, for example, you want to be 99 percent certain of the potential losses of a given risky event, then by calculating 2.58 standard deviations from the mean of a data distribution, you can determine what the maximum amount of losses you can expect with 99 percent certainty. In other words, you can be 99 percent certain that your losses will not surpass a given value. There are multiple implications of this. First, once you know the maximum amount of losses you can sustain and still successfully complete a mission, then these distributions allow you to estimate the maximum probability of sustaining them. Second, it allows you to manage the amount of total risk incurred between multiple operations in a combat theater, or multiple decision in a single mission, so that the total risk does not exceed a given value, or that any single item does not exceed a given risk value. VaR, in its original format, has some limitations which critics say lead people to incur greater risk than necessary. There are a number of variation on the core VaR model that improve on this, each with respective pros and cons, such as entropic VaR, conditional ARCH models, weighted models, and so forth.

Another method of calculating risk looks at the confidence intervals from the reverse direction. Rather than estimating how confident you are that losses will not exceed a given value, as with VaR and its various incarnations, expected shortfall estimates how bad things can get in the worst-case scenario. In other words, whereas VaR calculates with 99 percent certainty that losses will not exceed estimated values, expected shortfall calculates the expected losses within that 1 percent range. This accounts for one of the shortcomings with VaR; for example, a combat theater with 99 percent VaR below $10 million could still include 1 percent VaR of $10 trillion, so expected shortfall seeks to manage the amount of maximum risk incurred.

Both VaR and expected shortfall rely on data distributions. There are three primary ways to acquire these distributions. First is to use actual data from historical occurrences of having encountered the same risk, particularly within a similar context. The more data that are collected, the more accurate the estimates will be as a result of sampling and the law of large numbers (the larger the sample, the more accurate it will be in its representation of the population). The second method is the variance-covariance method, which utilizes a normal distribution with an adjustment made for the percentage of deviation of actual values from expected values at a given confidence interval. This does not require extensive historical data, but detailed collection of the data associated with the current conflict. The final method is called the Monte Carlo method, which requires you to run a huge volume of random simulation of potential outcomes and estimate the probability of losses in the same to calculate the probability of expected losses. If possible, it is greatly beneficial to do more than one method and compare the results, to help calculate the amount of deviation between them.

With huge volumes of data regarding a particular type of risk, there is also the potential to generate multicorrelative models estimating both the likelihood of such an event being encountered, and the volume of resources at risk. This is based on variations on the same types of calculations described in chapter 8, and requires data mining methods to identify those variables that have a significant correlation, and how to incorporate those variables into a predictive model.

The strength of each of these variables on the likelihood of a given threat being encountered and the severity of potential losses on those resources exposed can be determined, and even if the necessary data on a current or future operation cannot be collected, at least those variables known to be most strongly threatening can be given particular attention to mitigate that risk. These models are notoriously difficult to develop, however, because under dynamic conditions there are so many variables to consider, and generating an accurate sample can require vast amounts of data, all without guarantee that a model will ever be successfully discovered. When such a predictive model is found, however, its functional usefulness can be immense.

Finally, once all the risk values are measured, the total value of each potential decision can be calculated. Start with the expected benefits of the potential option of a given decision, then subtract the costs associated with pursuing that option, then subtract all the risk costs. The result is the net value of that option (discussed in greater detail in chapter 15). Recall that at the start of this process, you were facing a decision and developed as many response options as you could imagine. With the net value of each option measured, each of those options can be objectively compared and prioritized.

Even now, though, a final decision is not made, because not even all the steps until this point account for decisions regarding the treatment of each risk. There are four broad treatments for each risk. You can take steps to avoid that risk by altering your operations so that the risk has a lower likelihood of being encountered. You can reduce the risk, by managing the external environment in a manner that prevents the risk from being realized, or reducing the amount of exposure or severity experienced. You can share the risk with others, limiting the amount of total risk incurred by any one unit or organization. Finally, you can simply retain the risk and deal with it directly, absorbing the costs associated with it. How to treat each risk will depend greatly on what you are attempting to accomplish, the types and costs of risks, comparative organizational strengths and weaknesses, the resources available, and the amount of risk aversion maintained by the person making the final decision. Risk aversion can be measured in a person by using data about their past decisions. A risk lover will accept levels of risk that increase at a faster rate than return $((\Delta risk/\Delta returns) > 1)$, while a risk hater will only accept increasing levels of risk so long as the promise of returns increases at a greater rate $((\Delta risk/\Delta returns) < 1)$. Someone who is risk neutral will seek to optimize the risk to return ratio, coming as close to $(\Delta risk/\Delta returns) = 1$ as possible.

Finally, once the decision options have been prioritized and the risk treatments for each assessed, then all the options can be compared and the best option chosen, and the implications of that option understood, with immense precision. The resource-based comparisons of relative strength can be adjusted to manage a running model of combat outcomes, allowing you to estimate the volume of additional resources necessary for a mission to be successful after accounting for the resource costs associated with risk calculations. This ensures that resources are used most efficiently—never so few that a mission fails, but not so over-strength that the efficiency of the war effort is threatened (asset management is discussed in greater detail in chapter 15). If one risk generates a worst-case scenario, altering the optimal decision for future risks, then it can be beneficial to prefabricate alternative plans, or ways to shift between plans to maintain lowest potential risk.

Any one decision is only one point in a collection of decisions made in each mission, with each mission functioning as a single element of a larger operation, and each operation accomplishing one of many goals in the overall combat theater. This means that there are a vast number of decisions being made from potential alternatives at every moment, composing a generalized collection, or portfolio, of decision that can be measured and managed as an entity of its own, as shown in figure 14.2.

The curve in figure 14.2 represents the maximum amount of gains for a given level of risk for each decision. Each decision can fall within that curve, but cannot exceed, and the goal is to determine the optimal balance of risk and rewards incurred by the overall portfolio of decisions at each level. Note that portfolio risk decreases and then increases again. This odd behavior is the result of diversification in operations. It is, of course, highly risky to only pursue one mission or one operation, or to allocate all your assets to a single narrow pursuit.

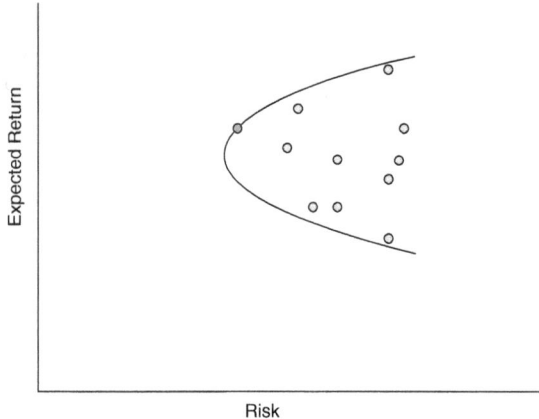

Figure 14.2 Operations portfolio

For example, when allocating infantry to the front lines of a traditional war, it would be extremely risky to put them all in a single city, rather than spreading them out over several cities. Assuming that allocating those resources chooses the best location first, then the second best, and then the third best, and so forth, and the cost (including risk cost) of obtaining and distributing those infantry resources slowly increases as one chooses the most efficient assets, then the next best, and so forth, the gains experienced as a ratio of the increasing rates of risk will diminish slowly at first, and then more quickly. At first, this diversification allows for decreased risk, before one's resources become stretched too thin, and the costs of pursuing each option is actually greater than the potential for added gains.

The optimal position in an operations portfolio is inversely proportional to the level of risk and the degree of risk aversion, and directly proportional to the level of gains. More risk is not necessarily a bad thing, if a person can consistently beat the odds and generate greater net returns on each decision than the probability estimates would predict. The success of the management of an operations portfolio can be measured using any of several ratios comparing the returns experienced to the amount of additional risk incurred. The Sharpe Ratio is one used in financial portfolio management that calculates the ratio of risk premium divided by the standard deviation of the excess returns.

The V2 Ratio improves on this by calculating the excess returns generated per unit of risk compared to some benchmark—generally an index of average risk to returns. The Sterling Ratio takes the net gains (recall: gains minus costs and risk costs) and divides that value by the largest single loss in the operations portfolio, as a novel estimate of risk losses.

These ratios evaluate the success of macro-level operational portfolio management by upper-level NCOs and officers at all levels. In more micro-scale assessments of the way risk was evaluated and managed, data collection is critically important to the future efficacy of risk management, and operational assessments can be performed using after-action reviews, collecting both quantitative data, and qualitative data that can be quantitatively incorporated into larger sets of metadata for use in improving future risk management efforts. The accuracy of risk assessments and success of risk management should be continually improving, the results of one decision contributing to improvements in the next, until the amount of risk to which you are exposed is entirely negligible, while exposing the opposition to the greatest amount possible. This is only possible by carefully understanding the nature of risk and collecting data that can be used to develop models and make decisions based on functional measures rather than subjective observational guessing.

CHAPTER 15

Asset Management

As we have seen in previous chapters, the resources that are available to each side in a conflict frequently play a major role in the outcomes of individual clashes and of the overall war. Despite the critical importance of this variable, though, it is still not as important as a determining factor in the outcome of a conflict as is the ability of each side to manage the assets that are available to them. Having huge volumes of resources available does little to contribute to one's success if one uses them ineffectively. If the volume of resources available was the only variable that mattered, then by all accounts the United States should have accomplished their goals in Afghanistan and returned home within days. Since 2002 (the first full year of operations after the attacks on September 11, 2001), US defense spending quickly rose and stayed consistently between 39 and 42 percent of the entire world's military expenditures—far more than the combined spending of the next 10–13 nations with highest global defense spending, but still could not predict the attacks on September 11, 2001. Granted, given the immense size of the US economy, defense spending remains only between 5 and 6 percent of GDP during that period, which is well within an average range compared to other nations, but that is not the point.

When the United States mobilized the full might of its primary and reserve defense infrastructure against a loosely associated collection of small militant groups in the Middle East who had nearly no resources available to them, often forced to rely on scavenged materials for their improvised operations, the United States suddenly became quite impotent. The primary difference was that the military insurgents of the Middle East were acutely aware of the potential of the scant few resources they were available to them, and knew precisely the best manner in which to use them. They knew that pursuing direct confrontation would accomplish little, and so they maximized the damage they could cause per unit of resources. They focused on hit-and-run operations such as roadside bombs, individual rocket attacks, and the ambush of small, isolated units. More than that, though, these organizations utilized their resources highly effectively in the manipulation of the local people and culture through psychological operations, which is something in which the United States failed miserably, shaping the opinions and activities of the region in a manner that prevented the United States from succeeding and leaving, so that they continued to remain exposed to the previously described small but persistent attacks. By contrast, from 2001 to 2012 the United States spent over $557 billion in war funding in Afghanistan, according to a 2011 report from the Congressional Research Service, and still failed to accomplish their stated goal. Clearly, the availability and consumption of resources matters far less than the manner in which those resources are managed and, to paraphrase sentiments of the US Chief of Staff of the Army's Strategic Studies Group, the US Department of Defense has become too big and complex to be effective—it is essentially tying itself into knots, incapable of efficiently using its resources.

Naturally, improvements in the utility of the resources available will be made by those methods described in chapter 12 ("Quality Management") and chapter 13 ("Efficiency Analyses"), but improvements are a relative construct. To accurately know the amount of benefit you are deriving from your assets, and maximize that benefit by improving the decisions you make regarding the assets in which you should invest, and how those assets should be allocated and used, you need to directly measure the assets and their potential usage to gain comparative absolute values. Throughout this chapter, analytics will be described, which will allow you to identify the most effective use of your assets, predict the best assets in which to invest, measure the speed with which you can change assets, and even assess the decisions and performance of those in charge of making the decisions.

First, it is necessary to clarify to what is being referred. Throughout this chapter, terms "assets" and "resources" will be used interchangeably. Both of them refer to anything available to an organization for use in its operations. In basic accounting, there is a simple equation that is used to illustrate the methods by which a company comes to own things: Assets = Debt + Equity. In other words, everything owned by an organization can be acquired either by incurring debt to purchase things, of by utilizing resources already owned by the organization or its owners. Another way to look at equity is that the value of everything owned by an organization not funded by debt belongs to the organization but can change form (such as using cash to purchase ammunition; both are assets but you can exchange one asset for another). Assets also refer to more than just physical items, since other things that can have great value for an organization include things like human resources (the people who work for/with an organization, and the unique knowledge and skills they bring with them), partnerships and alliances, and intangible assets such as information, methods, strategies, and even the image of the organization in the minds of the public. All these things have value, and they can all be acquired by using either debt or equity, both of which have costs associated with them.

Debt, in its most common form, refers to taking a loan and repaying it at a later date with interest. Under other circumstances it can also refer to any form of quid pro quo arrangement in which one party must perform services at a later date in exchange for goods or services provided in the short-run. For example, in exchange for the assistance provided by the people of an Afghan village in providing information, the village elders may require that the military forces provide them the cooperation of several soldiers to aid with defense against insurgents in the region who will target them for providing the requested information. Equity also has costs associated with it, such as the resources consumed by the construction of infrastructure, or the work required to increase strategic influence through psychological operations.

Sometimes, acquiring assets requires one to sell their equity, as well. To gain an ally willing to provide military forces of their own, they may require access to resources, or a degree of cooperation in decision-making that limits you from acting unilaterally. By deconstructing an organization into its component parts, it becomes clear that there is also a cost for each component in acquiring assets; in order to move people and resources to the front-lines, resources must be consumed by logistics and transportation units, for example. It is a cliché among economists to say, "There's no such thing as a free lunch," and for a military force to accomplish their goals they must utilize their assets, and those assets have a cost associated with them. When accounting for these costs of capital by subtracting their value from the total value of the goals achieved or the resources acquired, we are left with a calculation that better represents the value of what we have accomplished. This is succinctly summarized in another cliché—you may have accomplished your mission, "but at what cost?" Rather than in literary absolutes, though, this question must be considered as a literal application of analysis: Did you actually do greater good than harm and, if so, did you do it more effectively than your opposition?

To determine whether you created more benefit than cost (i.e., more good than harm), is answered most simply using a calculation called the breakeven point. Recall from chapter 14 ("Risk Management") that the risks to which operations are exposed can be approached as resource costs. The breakeven point, then, is calculated as the total resource costs over a set period of time or of accomplishing a goal, plus the resource costs associated with risk. If the amount of benefit generated by operations or by accomplishing a specified goal is equal to the total cost calculations, then the organization is said to have broken-even. Any operation that exceeds the breakeven point is beneficial, while operations that fail to reach the breakeven point were more detrimental to the overall operations of your forces than they were beneficial. Operations that meet their breakeven point, having 0 net gain or loss, are generally considered to be a bad thing as a result of something called the opportunity cost.

It is not enough that your operations should have a net gain, if your opportunity cost is higher than the amount of gain achieved.

The opportunity cost of investing in an asset or operation is the net value of the forgone option. In other words, if you have 100 units worth of resources, and 2 potential options for allocating those resources to specific missions, then those resources cannot be allocated to both places at the same time—you must choose. If option 1 has a net gain of 10 percent, and option 2 has a net gain of 5 percent, then the opportunity cost of choosing option 1 is the net gain of option 2–5 percent. When you subtract the opportunity cost from option 1, then you still have a comparative gain of 5 percent. Should you have failed to do your asset management analysis properly and picked option 2, then you would have an opportunity cost of 10 percent, and a comparative loss of 5 percent. Choosing the best option available to you, particularly under intense circumstances wherein the consequences of picking wrong are extremely high, can mean a world of difference. Even if you choose to invest in assets or operations that generates net gain, if it is not the best option then you are increasing the amount of resources that must be used to accomplish an equivalent goal, increasing the amount of risk to which people are exposed, increasing the amount of time it takes to accomplish the goal, decreasing efficiency, increasing room for mistakes, and generally limiting the strategic success of your operations. You have become less effective than is otherwise possible, and when an organization that is managing millions of people and billions of dollars in assets allocated using thousands of decisions each day, if asset management analytics are not being performed, then the impact this has could be enormous. If you are in a no-win situation, wherein each option will generate a loss, then by carefully analyzing which option will generate the least amount of loss could determine whether your organization will stop future losses, or fail completely.

One of the most simple asset-management calculations to perform is called Return on Investment (ROI). This is calculated by starting with the value of the investment, then subtracting the costs, taking your answer and dividing it by the costs (V– C/C). This allows you to more easily compare different types of investments by calculating the percentage of gain or loss, rather than absolute values. A particular investment might decimate $5 million worth of the opposition's assets, and cost a total of $3 million in asset, while another investment destroys $500,000 worth of the opposition and cost $200,000 in assets. Although the first option is creating $1.7 million more in net gain than the other option, the amount of gain per unit of resources consumed tells a different story. The first option provides a 66.7 percent ROI, while the second option provides a 150 percent ROI. Since the second option operates at roughly 10 percent the scale of the first ($500,000 rather than $5 million), if you could perform the second option 10 times, you would make much more progress. Another way to look at it is, by pursuing the smaller investment with the higher ROI, there will be a higher percentage of resources remaining to pursue additional investments. As a result, it is necessary to determine how many resources you have to allocate, then start with the investment that has the highest ROI, then the second highest, and so forth. If the total ROI of the smaller investments is still higher than the ROI of the big investment, then that is the best option to choose. If only a portion of the small investments have a higher ROI, so that the big investment has a larger total ROI, then choosing the big investment is probably the best option. As discussed in chapter 14, it is also necessary to consider the amount of risk inherent in focusing all your resources into one investment compared to diversifying your investments to limit the amount of potential losses incurred should that one investment fail.

Another calculation that can be done is called Return on Assets (ROA). Rather than calculating the amount of returns generated by a single investment, as with ROI, ROA calculates the average amount of assets available to the organization. This is used to measure the ability of an organization to make the most effective use of their resources, and is calculated by using the following:

ROA = (Net Gain/Total Gain) × (Total Gain/Average Total Assets)

This provides several pieces of information used in combination with each other. The ratio of net gain to total gain provides a calculation of something called the margin, which is the percentage of total gain achieved that contributes to the total value of the operations.

The ratio of total gain to average total assets ((Total Assets + Total Assets from Previous Period)/2) is a calculation of how effectively, on average, the organization is at utilizing their assets to further their cause. Recall that when multiplying two fractions if the denominator of one fraction is the same as the numerator in the other fraction that you can cancel that value, allowing you to now generate a ratio of net gain to average total assets, which calculates how effectively you are, on average, at utilizing your assets to increase the value of your operations.

An equation similar to ROA measures how effectively you are utilizing ownership value in theater operations not leveraged using debt: Return on Equity (ROE). A company called DuPont took ROE and broke it into several component parts so that, like ROA, it provides multiple ratios that each provide unique insight into how effectively assets are being managed—or will be managed, when used as a decision-making assessment. It is calculated as follows:

ROE = (Net Gain/Total Gain) × (Total Gain/Assets) × (Assets/Equity)

The first of these three, just as in ROA, is called the margin. The second is slightly different than in ROA, since it calculates the total gain generated from assets during the same period, rather than average assets, and this is called Asset Turnover. The final portion, called the equity multiplier, calculates the ratio of assets to equity, which functions as a measure of how effectively ownership in the operations not leveraged by debt is being utilized. Using the properties of multiplying fractions, that provides three more ratios:

(Net Gain/Assets): Measures how effectively assets were used to increase operational value

(Total Gain/Equity): Measures how effectively equity is utilized to increase competitive dominance through expanded operations that do not require later repayment

(Net Gain/Equity): Measures how effectively equity is utilized in increasing operational value

These calculations, together, allow for a rather insightful look into the organization's use and management of equity-based resources, without requiring context or prerequisite mathematical knowledge beyond basic arithmetic.

All these calculations are known as spot ratios—they measure a single point in time, and they allow one to make decisions based on predicted performance of investments, assets, and equity, or to generate hind-sight evaluations of performance. They do not, however, allow one to perform continuous assessments of actual performance compared to expected performance, or to account for the time-value of operations, such as is common in project management. In project management, the end value of a project is known, so as progress is made toward completing the project, one can measure how much value has been added to the project at each interval and compare that to the amount of value that should have been added to that project in order to estimate variation from both deadline and budget. By looking at each project as an investment, and pursuing multiple project, the operations of an organization are managed in the same manner as an investment portfolio. The same is true with military missions, operations, and strategies, which make it prudent to know how to calculate future value, present value, and net present value of these functions.

The future value (FV) of an investment or project is the known value that it will have at some future date, usually at the end of the project or when the investment fulfills its total intended purpose. The core calculation of future value is just that used for determining simple interest:

$FV = PV(1 + r)^t$

By starting with the present value (PV) of the thing being measured, and adding additional value to it at a specified rate (r) over a period of time (t), the value of that thing at any future date can be calculated. There are times when the added value can be reinvested so that it helps to generate even more value, called compounding. This requires only a slight variation on the core equation:

$FV = PV(1+[r/n])^{nt}$

Note that, like simple interest, the future value is a very simple function of present value, the rate at which value increases, and the period of time over which value increases, but in this you are also accounting for the number of times in that period that the value is compounded (n), referring to the number of times that the value added to the present value is reinvested and generates new value at the same rate as the original investment. If you happen to encounter a situation wherein you are able to continuously reinvest the additional value generated, such as through some forms of psychological operations (discussed in more detail in the Afterword), then this makes use of the exponential base, e, which is roughly equivalent to 2.72—a value that is used to accurately calculate things like population growth and, in this case, continuously compounding interest:

$$FV = PVe^{rt}$$

The inverse of these future value calculations provides an estimate of the value that something has now. When the value that an investment or project will have when it is complete is known, this calculation allows you to determine what the value is at any given point during the completion process:

$$PV = FV/[(1 + r)^t]$$

It uses the same components as the calculation of FV, except that it starts with the known future value and effectively removes value from it at a constant rate, known as the discount rate, over a specified period of time. The amount of value removed from FV is said to have been discounted. This calculation is critical when controlling operations, because it allows you to compare the actual current value of an investment or project to the expected value required by budget and deadline. If the actual amount of progress at time, t, differs from calculations of PV at that point in time, then the reason for this must be investigated. If the value is lower, then something is causing delays in progress, or unexpected rates of resource consumption. If the value is higher, then resources are being utilized inefficiently and could have been invested in additional projects.

Of course, under practical circumstances, the amount of value generated from a single project will most likely be accrued at multiple points in the future, and there will be multiple projects running at the same time. In order to generate a total calculation of all the value accrued by a single project, or even by multiple projects, a slightly more involved equation is used, called Net Present Value (NPV):

$$NPV = \Sigma PV$$

The simplicity of this can be deceptive. As used throughout this book, sigma, Σ, simply means to sum everything. So, NPV requires you to add together all the relevant present values. To help illustrate this, it is beneficial to break it down as a simple example:

$$NPV = 1000/[(1 + 0.01)^1] + 1000/[(1 + 0.01)^2] + 1000/[(1 + 0.01)^3]$$

In the example, a three-year mission will be divided between three equivalent units, each generating the same amount of value over the One-year deployments. Each unit will generate $1,000 in value over the year (remember that this is a very simple example). So, the present value of the first unit's contributions is $100. The present value of the second unit's contributions, since they will occur at a later date, is smaller than the first, at only $980.3. The final unit's contributions are $970.6, ending the mission. When adding together the total contribution of each unit on this mission, then, the mission's NPV at the beginning of the first year is about $2,050.9, which is discounted from the total future value of $3,000. If, at the end of the second year, the mission is much nearer to completion than originally expected, as measured by determining the actual NPV and comparing it to the expected NPV at that point, then some of the resources allocated to this mission could have generated greater value by being allocated to a different mission, thereby increasing ROA.

The careful allocation of assets is managed through a combination of calculating asset valuations, and a process called budgeting. When people think of a budget, they generally imagine a bureaucratic plan for setting restrictions on resources that is systematically violated. Far too often, this becomes a self-fulfilling prophecy because people end up using a budget for what they expect it to be used for. In reality, a budget is intended to act as a guideline to help you keep track of resource usage so that you can more effectively allocate resources, understand their rate of usage, and more accurately plan resource requirements. First, it is helpful to assess the costs of capital, most commonly utilizing something called the Weighted-Average Cost of Capital (WACC). This is a method of calculating the costs of capital associated with each source of capital, while emphasizing each as a proportion of the ratio of the total capital each represents. In other words, multiply the cost of your debt by the percentage of your capital funded by debt, and then do the same for equity. Formally:

$$WACC = [(D/Total\ Capital) \times C_D] + [(E/Total\ Capital) \times C_E]$$

where D = Debt, E = Equity, and C = Cost.

This can be broken-down into as many different sources and costs of capital as is necessary to generate an accurate calculation. The point of this is to estimate the average cost of capital for use in budgeting. This gives you a starting point for estimates of future costs of capital, which can be used in any capital-based projections, such as most of the calculations utilized in this chapter. It also provides a basis for time-series comparison, wherein you can track changes in the average cost of capital to identify increases or decreases in the efficiency of its acquisition and use.

Another calculation helpful in the budgeting of assets is the modified internal rate of return (MIRR), but before getting to the modified part of it, it is necessary to understand the core IRR portion. The internal rate of return functions just like NPV, with one difference: Instead of calculating the present value by discounting the rate of return, IRR calculates the discount rate necessary to make NPV equal to 0. In other words, what is the discount rate when NPV set to 0? This runs into a problem, though, in that both the cost of capital and the rate of return vary. Organizations, for obvious reasons, choose the source of capital that is cheapest first, then second cheapest, and so forth. They also invest in projects that have the highest returns first, then the second highest returns, and so forth. So, MIRR effectively does the same thing except that it takes into account any compounding reinvestments made, which vary in their discount rate from the initial investment and, if necessary, any additional financing costs that vary from the average cost of capital. In estimations of MIRR, the costs associated with financing capital for a given project may differ somewhat from the projected costs. With a lack of known capital costs, information from your WACC calculations can be utilized.

The actual allocations of resources to various projects include a variety of cash flows, both positive and negative. There are costs that are incurred as a result of the operations necessary to facilitate the project in addition to the resources utilized in the project, itself. There will be the immediate resource needs, of course, and the predicted core resources, and also something called working capital. Working capital includes any resources kept on-hand in excess of those actually predicted for use in order to manage daily, weekly, monthly, or other short-term resource needs, which can fluctuate before the predicted project completion. The management of short-term resource needs is extremely important. As briefly mentioned in chapter 14 ("Risk Management"), just because you have asset value doesn't mean those assets are necessarily in a form you can readily utilize. Sometimes you need to alter capacity of the current classes of assets, which provides the definition for economic short term and long term. Short term includes any period in which any give asset or operations cannot change capacity, as formally defined as the ability to replace any given asset. Whatever is the biggest, most difficult thing you have to replace, the length of time it takes to replace it defines your short term. The implication of this is that, in the long run, an organization can change the capacity of any of its functions or operations. Short-term asset management is concerned largely with measures of liquidity.

Liquidity refers specifically to the ability to turn assets into cash. Cash is the most liquid asset because it is already in a cash form. Inventories generally tend to be very liquid, as well. Things like real property, facilities, and huge equipment (concisely referred to as PPE: property, plant, and equipment) tend to take a much longer time. Why would a military care about its ability to turn its resources into cash? The same reason any organization cares: Cash buys other resources. Whatever you need, even if it is just to pay the bills during a time when continued funding from the government seems uncertain such as during a shut-down, cash can take care of it. The ability to do acquire cash is called solvency—when you have become insolvent, it means you have run out of cash, and can no longer fund your primary operations. That is a bad thing, so we use a handful of ratios to continue to effectively manage the level of liquidity.

Current Ratio: Measures as Current Assets/Current Liabilities. The word "current" refers to anything that will be transacted within 12 months. So, a current asset includes any assets that be turned into cash in a 12-month period, and current liabilities include any debts or incurred costs that are due within the next 12 months. This measures the ability of an organization to effectively fund its operations for another year.

Quick Ratio: (Cash + Equivalents + Receivables)/Current Liabilities. Cash is an obvious asset, but the others may not be. Cash equivalents include short-term money market investments, which government agencies frequently hold, as well as nongovernment entities. Receivables include any resources that were already due, and on which you are awaiting payment. The quick ratio tests the ability of the organization to continue to fund operations in case of more immediate fluctuations in budgeting that can sometimes occur in a more stable medium-term budget.

Inventory Turnover: Resources consumed in primary operations/Average annual inventory. Average annual inventory is calculated by using the current period's inventory, adding it to the previous period's inventory, then dividing the sum by 2—so it is a mean average. Inventory turnover effectively measures the number of times the total inventory is consumed and replaced in a period, and provides important information about maintaining minimum levels of inventory. As noted in chapter 13 ("Efficiency Analyses"), though, this can be a careful balance, since maintaining unnecessarily high levels of inventory will not only increase your operating costs, but also exposes additional resources to risk.

At the end of each period, there will a remaining balance on the budget. If it is positive, it means you have resources left-over, and if it is negative, it means you used more resources than you had in the budget. Many organizations have different ways of managing this—some allow the difference to "roll-over" into the next period, defining the next period's starting budget. Others require "use-it-or-lose-it" resources, wherein the resources are reallocated to the primary resource pool and taken from individual operations. It is the absolute insanity of the US government that they take this one step further. If a particular agency or operation does not utilize its full budget for a period, then their budget is decreased the next period. This fails to allow agencies to account for long-term planning, and encourages tremendous amounts of waste as everyone rushes to spend whatever is remaining on their budget at the end of each fiscal year on nonsense, just so they do not lose their budget next period. This has a tremendously harmful impact on asset management and efficiency, and significantly slows operational growth.

Sustainable growth can be looked at in two ways. First, how quickly can a project grow by remaining internally funded? Internally funded growth is an important concern in that it provides critical budgeting and asset management information about the rate of growth a project will experience without requiring additional sources of capital other than those generated by its primary operations. Particularly, in a highly competitive or risky environment, where incurring debt in any form may increase costs beyond what is deemed acceptable. Sustainable growth in this respect is calculated by dividing net gain and dividing that by their total assets, or total equity, depending on preference of the treatment of a project that may be already heavily leveraged in debt. The result of this calculation is a ratio of how effectively a project is utilizing their existing assets to generate value growth, which helps to estimate future growth in resource requirements. Another look at the sustainability of growing operations, though, is to consider the manner in which an organization functions within its operating environment. No, this is not about environmentalism, it is about the rate at which resources are consumed, and the rate at which those resources can be replenished and made available for purchase and use again. If you are consuming any resource faster than it can be replenished, then that resource will eventually be depleted (which, by itself, has been a strategy of warfare throughout history, to ensure the opposition cannot access certain key resources). When consumption rates are higher than replenishment rates, you have reached a point called degradation levels of consumption. When replenishment rates exceed consumption rates, you have reached a point called restoration levels of consumption, and this isn't efficient, either. At degradation levels of consumption, the resources will deplete, become more expensive, and less available. At restoration levels of consumption, the resources aren't being utilized as efficiently as they could—you are not making full use of your potential sustainable growth rate. Although restoration levels are considered sustainable, the optimal rate of consumption is one which perfectly matches the restoration rate, called static-state consumption.

When you incorporate risk into the data for asset management, it is possible to generate extremely powerful asset management models. Remember from chapter 14 ("Risk Management") that Beta is a measure of volatility, and that much of the existing literature still utilizes Beta as a risk metric despite lacking validity (primarily because that's the best they could come up with). Ignoring that for a moment, and pretending Beta is merely a measure of real risk, and not volatility, the core CAPM model still has value in illustrating some key points.

CAPM: $r_s = r_f + \beta(r_m - r_f)$

where R_s = return on a specific pursuit, R_f = risk free rate of return, and R_m = market returns

This is the Capital Asset Pricing Model, which illustrates a very simple, but not functionally useful, valuation calculation of a single investment or project. The risk-free rate is the benefits gained by pursuing the course of least risk, rather than least benefit. So, the amount of benefits that must be gained by incurring any additional risk, then, must be greater than the risk free rate in a proportion at least equal to the value calculated as risk multiplied by average operational returns in excess of the risk-free rate. In other words, this helps to illustrate how much value a given operation must be offering in order to make it worth pursuing. Another way to look at it is through the calculation of alpha:

$$\alpha = r_s - r_f + \beta(r_m - r_f)$$

Rather than calculating the estimated value of the project, alpha calculates whether the risk-adjusted returns are high enough by subtracting the CAPM model from the specific returns. The implication is that if alpha is 0, then the project has generated exactly enough returns to be worth pursuing. If alpha exceeds 0, then it has generated something known as excess returns, which is a good thing. If returns were less than 0, then the project was a failure. This model uses several assumptions that make it fail in functional usefulness, though, so it has been replaced by one called APT.

Arbitrage pricing theory utilizes a similar basic relationship between variability and relative returns as a way of determining the value that a project must have to be worth pursuing, but it is much more adaptable.

$$\text{APT: } r_s = r_f + \beta_1 r_1 + \beta_2 r_2 + \cdots + \beta_n r_n + \varepsilon$$

Rather than being constrained by specific contributors to returns, and rather than assuming all factors adjust perfectly and immediately to changes in their environment, and rather than assuming Beta measures risk, APT functions as a create-your-own valuation model by providing a basic core and then forcing you to customize it. The ability to adapt the model to any type of situation, and the exact accuracy of the model, then, is very dependent on the data you have available, and your personal ability to manipulate it. Once again, the valuation starts with the risk-free rate, but then jumps into a potentially limitless series of other variables, each of which generate contributions to the returns on a project, and the variability of those returns. So, if you wanted to incorporate a risk variable as a cost to returns, you could include it as a negative value that generates negative returns, and the more certain you are to encountering that risk, the lower the variability would be. The epsilon at the end of the model is an error function—recall that statistics relies on the ratio of effect-to-error, or the amount of variability that can be explained utilizing known variables, and the amount for which the model cannot account. The epsilon is a value included to represent the amount of variability for which the remaining variables cannot account.

Once all the analytics necessary to estimate and compare the values and risks of each project is performed, and the statistical likelihood of success is calculated based on the strategic considerations of basic operating analyses, now it is possible to develop strategy based on all of it. Strategy is nothing more than a series of decisions, as illustrated in figure 15.1.

The figure is called a decision tree, and each open circle on it is called a decision node. Each decision node represents a point at which you have several potential responses, each response is represented by the pathways stemming from the node. Most pathways lead to a new decision node, while others lead to an end node, which indicates the end of the strategy. It is possible for several pathways to converge onto a single decision node, which indicates that several potential decisions bring the strategy to the same point, but by different methods. The best decisions in the strategy are those that create the most value, but several must be available in a dynamic environment in case events make a decision no longer viable, providing secondary and tertiary decisions to the same conclusion. The value of each pathway is calculated by utilizing the net values estimated using equations that include gains, costs, and risks, such as the APT model. That net value is then multiplied by the estimated probability of success of the decision, as calculated using predictive and probability modeling discussed throughout

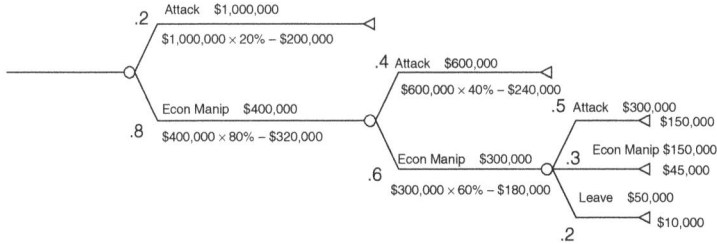

Figure 15.1 Decision tree

part II of this book. By multiplying the net value by the probability of success, the value of each decision is calculated, and the best decision can be chosen. The path from start to any end that generates the greatest value is considered the most valid strategy. Then, as any conflict is composed of multiple strategies employed simultaneously, the value of each strategy can be input into the same basic portfolio graphs and models discussed in chapter 14 ("Risk Management") to assess how effectively strategies are being managed, and the success of the person managing those strategies. In the end, the entire conflict is won even before it has begun.

> The one who figures on victory at headquarters before even doing battle is the one who has the most strategic factors on his side. Observing the matter in this way, I can see who will win and who will lose.
>
> Sun Tzu, *The Art of War*

CHAPTER 16

Challenges and Limitations

Each of the many operational analytics discussed contribute to refining decisions and improving every function within an organization. As each one is performed, one after the next, new information and ideas will be generated, and inferior options eliminated, gradually revealing the optimal path. Understanding the methods by which goals will best be achieved, though, does nothing to alter the limitations of an organization to perform. Even so, the ability to minimize harm in a no-win scenario, or be as effective as one's resources will allow, will still be critical to the survival and success of the overall mission, and this is only possible once the necessary analytics have been performed.

There is a large human factor involved in ensuring that due diligence is performed and that actions are taken, which are consistent with the data. Far too frequently, decisions are made for reasons other than optimizing operations, such as personal gain, pride, denial, nepotism, and so forth. In addition, discovery analyses cannot directly reveal all potential options, only guide the analysts to them. It is up to the analysts to utilize brainstorming methods, systems mapping, and a variety of analysis tools to identify all the potential options; they must be creative, and be supported by an environment that supports creativity. Foreseeing various types of risk comes from a deep understanding of the circumstances in which the risk is presented, requiring experience, or at least the past experience of others collected and compiled into a database, which can search for common themes in the internal and external environments. The ability to manipulate the data in such a way as to make meaningful and accurate information available is limited to knowledge of mathematics and computer science. In the end, an analysis is only as good as the analyst, and even a good analyst is frequently not the one who makes decisions, but rather functions as a sort of consultant providing input, which is often ignored.

As with all forms of statistical analysis, the quality of the analysis will be limited by sampling methods, available data, and the potential for statistical error, no matter how small the potential. As was discussed in parts I and II, it is in the nature of statistical analysis that data will be imperfect due to limitations, which is why sampling exists. Sampling itself, though, also has limitations, which only allow us to say within a given level of confidence that a particular statement is true. It is in the nature of research that the findings and their contributions to theory are based on the best knowledge available, and that there will always be the possibility that future knowledge could prove the current knowledge obsolete, no matter how remote that possibility seems. It only takes one study to debunk centuries or more of what we think we know, but then we must adapt and utilize the new knowledge to improve. The cycle becomes a never-ending evolution of knowledge providing the foundation for improved methods and technologies, which keeps forces on the cutting-edge of strategic advantages at an ever-increasing rate.

The limitations of operational analytics can, in themselves, by analyzed to determine the degree to which they allow error and inefficiency by studying them in hindsight. This allows for the ability to estimate exactly how much they should be relied on, and whether there are consistent patterns for which plans and operations can adjust. That which is not known can sometimes provide just as much information as that which is known, if used properly. The amount of error in a model can be incorporated into correlative models to determine which variables are creating the most error. Just because the knowledge is available, though, does not mean that people can be forced to use it.

CHAPTER 17

Suggestions for Future Research

Operational analytics is nothing more than an applied form of other, generalized forms of analytics. All the chapters described in part III directly utilize the same descriptive and predictive analytics discussed in parts I and II, and applies them to functional operations of any organization. In most cases, however, the analytics being utilized in operations are not exclusively one type of calculation, as discussed in other chapters—they tend to incorporate one or more types of analysis into a more complex model that describes the nature of the relationship between an organizations' operations and the environment in which that organization operates. So, many of the potential direction for research related to operational analytics focuses on the individual types of analytics being utilized, but on how to best use each, and how to integrate them into cohesive applications.

One of the more complex relationships is that of risk and reward. The phrase, "no risk, no reward" is a misguided misnomer that often leads the speaker to accept unnecessary levels of risk. The nature of this relationship is that for an individual to accept the additional risk, there must be additional reward equivalent to or greater than the incurred risk. Despite this very simple piece of logic, the models used to calculate both risk and reward are muddled and often uncertain. Even professionals sometimes make the mistake of using measures of volatility as a measure of risk, while true calculations of risk can be extremely difficult to calculate given the sometimes very large number of variables that play a role in the success or failure of a mission. Even the expected gains from a particular operation are not always clear in the beginning, and statistical probabilities of what will be accomplished are used, as well as probabilities of the amount of resources that will be consumed in the process. There is a vast volume of research on these topics in the field of finance—investing firms spend huge volumes of resources to develop algorithms that automatically execute specific instructions when specific flags in price and volume are raised. Not only can much be learned from these algorithms when applied to other fields, but so can the response to these algorithms. In an example of how one's actions can trigger a chain-reaction of responses that fundamentally alter the operating environment, and how that can be taken advantage of, something exists called a flash crash. A huge proportion of the total stock market is managed by professional investing firms such as banks, hedge funds, mutual funds, broker-dealers, and so forth, and most of them are utilizing automated algorithms to manage their operations. A flash crash occurs when one significant trader takes an action—say, sells their investment—which causes a change in the environment that triggers many others to respond—for example, the lower price caused by the initial sale of an investment may trigger everyone else's systems to sell-off their investments, too. This fundamentally changes the pricing of the entire market for just a few moments, before the lower prices trigger the bulk of everyone's systems to repurchase at the lower price. The person who can take advantage of those lower prices before the rest will generate a significant amount of resources for himself, and this all happens over the course of no more than a few minutes, or even seconds. The organization that wins is the one that utilizes the best analytics.

Finally, the ability to execute operations with speed and precision depends entirely on preparation. As is often said in the US military, "prior planning prevents piss poor performance," and one of the most important ways to ensure the success of operational analytics will be the use of prefabricated models and formulas capable of repeated success in specific simulations. Collecting data on practically everything will be critical to the accuracy and effectiveness of these analyses, ensuring that the data required to develop statistical and probability models is available. Then, the systematic digital organization and categorization of this data is performed so that it can be easily and relevantly accessed and analyzed, and that conditions being encountered can be cross-references with similar conditions of the past. Stress tests, to discover the limits of the efficacy of the data under varying conditions will also be necessary to understand their applicability. Much of the research associated with correlative analysis and statistical modeling is associated with data mining, and the simple identification of relevant variables and the influence they have. It can be a long and tedious process, which frequently yields no new information, but this is one type of work in which military agencies tend to excel. With huge volumes of people who are often left idle, "extra duties" are frequently assigned as busy work, "hurry up and wait" becomes an unofficial motto, and sometimes little gets done at all. With improved efficiency, the amount of downtime throughout the entirety of HR will be largely eliminated, allowing for a reallocation of HR assets to productive research. The information derived from these efforts can then be used to improve the accuracy and reliability of current models, develop new models, and even make a wealth of new discoveries. The term "big data" has swept the private sector with a rarely seen fervor, and with good reason, but "big data" refers to nothing more than utilizing many of the methods described throughout this book.

Afterword

This book is one in a collection of three, which I know as the Modern Warfare Collection. The other two books are *Psychology and Modern Warfare*, and *Economics and Modern Warfare*, and together the three of them overlap in several places, one enhancing another, and in the ways each is unique the strengths of one supplements the weaknesses of another in addressing the challenges of executing a comprehensive strategy. In combination with information warfare, also known as computer warfare, these things represent the future of warfare, law enforcement, political and social influence, and business strategy. There will not be a Modern Warfare book on information warfare because I'm simply not capable of writing one and, even if I were, there are authors far more competent in the topic who have already written about it. Conflict comes from differences in idea—ideas about who should run things, the way things should be run, ideas about resource rights, about fairness and the socioeconomic disparity, about taxation, about cultural and idealistic differences, and about religious manipulation. Despite all the efforts of those who dedicate their lives and resources to finding a way to find peace by ending conflict, this approach is destined to fail so long as people hold different ideas about the nature of the way things should be. The way to resolve warfare is not to end conflict, but to make traditional militaries obsolete by providing a more effective manner of accomplishing their goals. This is found in a comprehensive approach to information operations, economic operations, psychological operations, and quantitative analytics.

From the perspective of analytics, the overlap exists with economics in the form of something called econometrics, and in psychology through psychometrics. These are fields of specialization in economics and psychology, respectively, which focus on quantifying variables, developing mathematical models, developing and tracking measurements, discovering relationships, predicting outcomes, and essentially doing everything listed in this book except focusing on their respective fields rather than on focusing on strategic applications.

Economic warfare refers to any of various methods and tools intended to alter resource distribution, and resource-based behaviors for strategic purposes. This was divided into supply manipulation, trade manipulation, and market manipulation. Supply manipulation focuses on increasing or decreasing the availability of supplies, human resources, and capital directly or via supply chains, and taking advantage of changes in actions and behaviors, which result. Trade manipulation focuses on altering the value of exchanges to increase or decrease the consumption of resources, or alter the value of resources. Market manipulation focuses on directly altering market forces to manage behaviors, or deriving information from markets that are used to extract intelligence about the combat theater.

Economics is primarily a mathematical field, so econometrics naturally becomes critical to the efficacy of economic warfare. Price, volume of supply, volume of demand, and the relative values in exchange are the broad, primary measurements utilized, and are applied in a variety of ways. For example, by measuring changes in the price and availability of specific things in a region, as well as the behavior of opposition supply lines and asset movement, it becomes possible to predict with a high degree of accuracy what they are planning to do and when they plan to do it. By carefully monitoring specific market and financial activities, it is possible to estimate the rate of resource consumption and develop strategies, which cause the opposition to consume their available resources at an unsustainably high rate. The strength of each side in a conflict is very much the result of the capital, resources, and human resources available to each side, which can be incorporated into statistical models to accurately predict the outcome of a specific battle, as well as the ability of the winner afterward to accomplish further missions.

Psychological warfare refers to any of various methods and tools intended to alter the mental and behavioral processes and patterns that people exhibit for strategic purposes. This, like the other books, was divided into three categories: idea modification, emotional modification, and behavioral modification. Idea modification focuses on methods that manage what people know, what they believe, and what they think. Emotional modification focuses on methods that alter how people feel about various groups and events, their emotional well-being, and their level of motivation to accomplish goals and support a particular cause. Behavioral modification focuses on methods that alter the behaviors people exhibit, the way in which they interact and organize themselves, and the decisions they make.

Psychology tends to be less quantitative in nature than economics, and focus more heavily on qualitative methods, however analytics are commonly utilized to measure the strength of interactions, the efficacy of interventions, and the probability of specific types of responses. For example, how do increases in the exposure to specific ideas change the rate at which those ideas are internalized by those exposed to them? What percentage of people with PTSD or other anxiety disorders can be cured through memory-alteration treatments? Answering these types of questions is critical to understanding how to utilize psychological methods.

As noted earlier in this chapter, computer warfare is also playing an increasingly important role in world events. The global reliance on computers for a huge variety of daily operations means not only that this poses potential to enhance or disrupt those operations remotely, but that important information can be collected about these operations, as well. Analytics play an important role in the ability to accomplish this, such as through programming, and engineering processes. The numerical analysis of algorithms is utilized to determine the amount of computer resources required to execute the algorithm. Just as importantly, however, performing analytics frequently requires the use of computers and custom programs or algorithms that can process huge volumes of data far more quickly than people.

All these things are being increasingly utilized in a variety of fields. Law enforcement is starting to realize the importance of analytics through predictive policing, and regularly use supply lines to find manufacturers of illegal narcotics and counterfeit currency. Hostage negotiations rely heavily on psychological methods. Political organizations rely extremely heavily on psychological methods to manage public opinion, and utilize economic influences as leverage in negotiations. Economics has much of its roots in business, while businesses have been increasingly relying on psychological methods in marketing and human resources since the 1930s, and analytics have recently gained a huge popularity through the field of business intelligence, or "big data." There has been a disparity in the books and coursework in the social sciences, and their application. The educational material on these topics focus on passive methods—responding to market forces, guiding people toward their own psychological treatments, and avoiding intervention in social structures as much as possible during research. The reality, though, is that active management of these things is a common practice in a variety of fields.

Military strategy is particularly useful as a method for studying these fields, however, because the limitations of these studies can be assessed through historical applications that would typically be considered unethical in normal practice. While not all the methods described in all these books are ethical in all applications, their use in defense applications holds the potential to make traditional military methods obsolete, which not only gives a moral imperative to pursue them in order to save lives and infrastructure, but also as a matter of improving our nation's defense capabilities beyond anything the world has ever seen. There is strong resistance to changing the status quo within the outdated and inefficient defense infrastructure, which relies heavily on its ties to heavy industry. As other industries around the world surpass the capabilities of the military, however, the defense industry will be forced to adapt if they want to stay relevant in a modern world, or else be replaced by private industry offering services more effective than their own. It is my hope that these books will act as a guide in the development of a private industry dedicated to improving business, law enforcement, politics, and, eventually, defense.

Critique of Current Methods

The importance of using analytics as a source of strategic advantage has not entirely escaped the defense industry, but their use has been inconsistent, often being dismissed in favor of reliance on personal experiences (also known as guessing). Modern research in analytics is not frequently performed directly by the defense industry at all, and such organizations as the Chief of Staff of the Army's Strategic Studies Group openly acknowledge their rejection of modern methods that are already successfully used in other industries due to past implementation failures. Rather, there are very tiny units in the navy and air force dedicated to this matter, while the bulk they sit idly and unaware, waiting while mathematicians, statisticians, computer programmers, and others who work on these problems as a labor of love try to generate solutions for improved methods in defense operations. Developments in defense analytics most commonly come from private companies attempting to sell products, recommendations made by think tanks trying to get research grants, or from university professors who intend to publish the results. Most of these analytics never find their way into defense operations at all and, of those that are adopted by the defense industry, it is difficult to say the exact extent to which they are utilized because once they reach that point, their use becomes a matter of top secret national security. So, any critique on the utilization of analytics in the defense industry will have something of a gap in it. It is possible to study what is being developed, what is being ignored or rejected, and the changing nature of operations as a possible indicator of the utilization of analytics (or lack thereof). It is even possible to assess the use of analytics in the broad scope of military operations, but most analytics developed focus on high-level decisions and the manipulation of intelligence data, putting them out of reach to those not directly involved in those functions.

The vast majority of analytics utilized in warfare are associated with the physical properties of equipment. The blast radius of a bomb in areas of varying population density is frequently used to estimate the number of casualties and the value of damage resulting from detonation. The properties of potential and kinetic energy and their influence on ships or the ballistics of firing a bullet are well established. The flight of a jet or a rocket and how to compensate for meteorological conditions is understood in great detail. The Army Corps of Engineers regularly analyze geography and the physical properties of building materials. Mechanical diagnostics of vehicles and weapons are daily practice performed with absolutely no problem. These things are all important considerations when developing strategy, but they do little to facilitate the development of strategy, or to improve operations. To accomplish this, various type of statistical analytics are required, which is something that is clearly lacking among military forces worldwide.

Military forces have attempted to harness the intent of some extremely useful analytics in a manner that does not actually utilize the analytics, themselves. The result is something that sounds useful, but does not provide any actual insight or new information with which decisions, operations, or strategies might be improved. The assessment for risk management currently utilized, called composite risk management (CRM), is nothing more than a casual, observational assessment with arbitrary labels that draws no conclusions, which were not already drawn based on the immediate response to the observation. In other words, CRM breaks-down the process by which people naturally assess risk into a set of formalized steps that provide no additional information not already available. Cycles of varying sorts are very popular, and present a nice philosophy on how decisions can be made and operations improved, but, once again, offer no methods for extracting additional information about a given situation, or tools with which to address it. None of these casual, observational, general-use assessments are actively implemented, as there is nothing to implement. Granted, the majority of military personnel are not familiar enough with statistics to perform applied analytics on their own, but with recent advances in computer technology and even telephone technology, none of them are beyond the capabilities of a phone application, or "app." It is these applied analytics, intended to accomplish specific things, which provide the kind of useful information and insights that lead to tangible improvements at all levels of an organization.

As noted earlier in this critique, the exact utilization of applied analytics is generally considered top secret, so little is known about it that can be shared publically in a book like this. Instead, an indirect method of assessing the efficacy of any possible analytics being utilized can be performed by looking not at the analyses themselves, but at the decisions and operations made which would, or rather *should*, utilize analyses. In looking at military analytics from this perspective, the picture that starts to form is one of complete underutilization, if any utilization of analytics exists, at all. The response that the US Department of Defense has to world events is purely political and reactionary. Shows of force and/or prowess are common in military parades and drills, but rarely do these have a point other than to posture in view of perceived threats. Intelligence is based entirely on direct observation and collection of information about plans, such as through satellite imagery, audio interception of conversations by plane or by tapping phones, or penetrating computer systems for the direct collection of opposition plans. None of this utilizes numerical analytics. Attacks are sometimes stopped when this direct intelligence can be useful, but most of the attacks are stopped purely by reactionary response. Never has any nation successfully used mathematical analysis to predict the timing or location of an attack, and even when direct, nonnumerical intelligence is available, as it was previous to the attacks on the United States on September 11, 2001, the response was still not enough to prevent them from happening. Even in operations it becomes clear that those making decisions are not certain what volume of resources will be needed to accomplish their goals, as the number of people and assets deployed to various regions are subject to large and frequent variations.

Estimates of the efficacy or success of operations or individual missions show little thought given to the context or influential variables on their execution. The sluggish and intensely inefficient operations of military forces demonstrate a clear lack of useful performance analysis. By ignoring our ability to describe, predict, and effectively respond to those things which are happening, and which are going to happen in the future, we are recklessly guessing, never truly knowing what we are doing or what the outcome will be. When the stakes are high, such as the life and death situations of warfare, this is simply not acceptable. For businesses, the stakes can be just as high, when one single bad strategic decision can mean the difference for the lives of thousands of people when they no longer have a job, or when the world's largest economy comes to a complete halt as a result of poor risk analytics performed in a single sector (finance), thereby devastating people globally.

In contrast to the lack of progress being made in the defense sector, a small percentage of people around the world have been making increasingly frequent advances in improving the analytics that are available for defense purposes. Think tanks such as the Institute for Defense Analysis, The Dupuy Institute, frequently work on quantitative analytical solutions. University professors and researchers often work on these issues, such as at the Krasnow Institute for Advanced Studies at George Mason University, MIT's Center for Collective Intelligence, University of Maryland (which includes such field experts as Thomas Schelling, and V.S. Subrahmanian), the West Point Network Science Center, and other academic researchers such as the renowned mathematician Lewis Fry Richardson, who was a leading voice in the early application of modern defense analytics as a way of promoting pacifist conflict resolution by making warfare obsolete. Private companies trying to sell products to the government are commonly working on these projects, as well, including such companies as Aptima, and Lockheed Martin. Branches of the government such as the NSA also hire internal developers to engineer software that can perform data collection and analysis.

Since September 11, 2001, however, much of the progress that has been made in utilizing defense and intelligence analytics has overstepped certain boundaries of privacy, constitutional amendments against illegal search and seizure, and against the ethical sensibilities of the people who these programs were intended to defend. The measures taken after the attacks on 9/11 did not change the world, but they did change US national policy, including expanding the use of mass-surveillance techniques that collected a wide range of data from a wide variety of sources. The Department of Homeland Security uses methods of automatically scanning for keywords in conversations which, when found, are investigated more thoroughly. Several data collection programs utilized by the NSA were revealed in an article published by *The Guardian,* which collected and analyzed huge volumes of data to pick-out patterns in the usage of particular types of words and phrases through something known as meta-data, which is data related to the data (in other words, instead of looking at the raw data of word usage, it might analyze the number of times specific types of words were being used in a region, or, in the case of Boundless Informant, track data on the sources of their data). Analysis requires data, and there is a very direct trade-off between the amount of data collected, and the usefulness of an analysis, so in order to make use of these analytics, the question becomes one of data collection methodologies – how can the required data be collected without violating privacy rights? The reality is that every major nation spies on each other, and on their own people; this is not a secret in the slightest, and the fact that they are doing this is written right into the laws of each nation. Occasionally, spy planes will get caught over one nation or another, or undercover agents will get caught, and the use of satellite imagery analysis is well known and available even to companies such as Google. During the administration of Bush II, it was revealed that the US government was illegally tapping phones without a warrant, and the scandal was short before the government retroactively made the activities legal. So, by the time the article in *The Guardian* was published, nothing that was revealed was new information, but outrage ensued because of the very public release of its information in a manner that presented the information as new, and the inability of the government to stem the media frenzy that comes with a scandal. The public already knew it, but became outraged anyway as a result of the manner in which the information was presented, in a simple example of the manipulation of public opinion (described in great detail in the book *Psychology and Modern Warfare*, in the same collection as the book you are currently reading). In reality, many of these data collection methods are quite harmless, and do not collect any identifying information, thereby protecting personal privacy. Not all of them, that is true, but many of them. Data collection, even for defense purposes, can be collected in an ethical manner by utilizing such nonidentifying data. The public, generally speaking, assumes much worse, so backlash is inevitable for any action, and becomes a matter of proactively utilizing public relations and marketing techniques.

Though there have been consistent advances in defense analytics throughout modern history, the actual rate with which they are successfully implemented is quite dismal, and this lack of emphasis results in such work receiving less attention than it should. Though the importance of this line of research is acknowledged by a handful of people in the defense sector, and a very quickly increasing number of people in the private sector, the implementation and application of these methods are far less than even a half-effort. The results of the lack of emphasis given are evident in the lack of efficacy and performance exhibited by the various militaries around the world, despite the vast amount of resources allocated to many of them.

Index

Abductive, 71, 93–5, 97, 101, 105
Absolute Advantage, 130, 131
Afghanistan, 99, 140–2, 153, 154
Albert Einstein, 14
Alexander the Great, 125, 137
Alpha, 168
Alternative Hypothesis, 8
Altman's Z-Score, 83
American Society for Quality, 115
ANCOVA, 42, 44
Anderson-Darling Test, 51
ANOVA, 31, 41–5, 86, 91
APT, 168–9
Aptima, 186
Asset Management, 76, 129, 149, 153–69
Assumption, 22, 25, 47–50, 52–4, 59, 66, 73, 92, 168
Attrition Risk, 143
Autocorrelation, 90
Average, 1, 15–17, 19–20, 22, 24, 31–2, 35, 37, 48, 86–7, 124, 126–8, 135, 142, 151, 153, 158–9, 163–5, 167

Bartlett's Test, 153
Bayes Theorem, 80
Bayesian Probability, 77, 80
Bell Curve, 15, 22, 31
Benchmark, 39, 123, 128, 136, 151
Bernoulli Trial, 78–9
Beta, 140, 167–8

Bias, 26, 29, 58, 67
Big Data, 3, 111, 175, 180
Binomial Distribution, 78–9
Blitzkrieg, 125
Bollinger Bands, 32
Bottleneck, 134
Boundless Informant, 187
Box Plot, 8–19, 31
Box's M, 53
Breusch-Pagan Test, 53
Business Intelligence, 2, 145, 180

Capital Productivity, 128
CAPM, 167–8
Carl Gauss, 77
Categorical Variable, 13, 42–4, 79
Census, 26
Center for Collective Intelligence, 186
Central Limit Theorem, 29, 77
China, 100
Cohen's d, 45
Cohen's F-Squared, 91
Comparative Advantage, 130–1
Comparative Assessments, 31, 37–46, 51–2, 79, 86, 90
Confidence Interval, 28, 116, 133, 147
Continuous Variable, 13, 38, 43
Correlation, 25, 30, 56–7, 63, 72, 80, 84–94, 121, 147
Covariance, 29–30, 42, 56, 89–90, 92, 140, 147

Covariate Space, 25
CQT, 119
CRM, 184
CRUSH, 95
Cumulative Distribution, 50
Current Ratio, 165
Curvilinear, 33–4

DARPA, 90
Data Diagnostics, 47–64, 66, 184
Data Mining, 52, 107, 147, 175
D-Day, 97–8
Decision Tree, 76, 169
Deconstruction, 112
Deductive, 71
Degrees of Freedom, 39–41, 45
Delta, 33
Density, 15, 18–19, 35, 52, 79, 95–6, 98–9, 184
Dependent Variable, 14, 32–3, 42–4, 67, 69–72, 80–1, 85–8, 90–1, 108
Derivative, 33
Descriptive Analytics, 4, 7–10, 12, 58, 61–2, 66, 107
Descriptive Statistics, 8, 11–22, 26, 31, 37, 49, 55, 66, 77
Deviation, 15, 18–24, 28–32, 37, 45, 48–52, 69, 77, 87, 116–18, 122–3, 146–7
Distributive Efficiency, 128–31
DMAIC, 118
DMPO, 116
DOWNTIME, 136
DuPont, 159
Durbin-Watson Coefficient, 55

Econometrics, 5, 178
Edward Altman, 88
Efficiency, 35, 123–37, 149, 154, 157, 163, 165–6, 172, 175
E-MEME, 94
EOQ, 134–5
EPIC, 94
Eta-Squared, 91

Europe, 100
Expected Shortfall, 147

FANOVA, 42
Fast-tracking, 133
FOB, 129
FOIA, 94
FV, 160–1
F-Value, 41

Gamma Distribution, 79
Gantt Chart, 133
Gaussian Distribution, 77
Geospatial Intelligence, 72, 81, 93–101, 105
Germany, 125
Glass's Delta, 45
Goldfeld-Quandt Test, 53
Greece, 125
Guinness, 38

Hammer and Anvil, 125
Hazard Risk, 143
Heat Map, 14
Henry Ford, 118, 136
Heteroscedastic, 52
Histogram, 31
Homogeneity, 52–3
Homoscedastic, 52–3
Human Resources, 43, 127, 155, 178, 180
Hypothesis, 8–9, 39, 44

Independent Variable, 14, 32–3, 42–4, 67, 69–72, 80–1, 85–8, 90–1, 108
India, 100
Inductive, 71
Inferential Statistics, 8, 26, 73
Institute For Defense Analysis, 186
Interquartile Range, 18, 20, 31, 55
Interval, 13, 18, 28, 79, 116, 133, 147, 160
Inventory Turnover, 135, 165
IRM, 141–2

IRR, 163
Isaac Newton, 84
ISO, 141

Japan, 100, 124
J-Curve, 35
JIT, 135
Joseph Mengele, 28
Judgment Sampling, 26

Kaldor-Hicks Efficiency, 130–1
Kanban, 135
Kandahar, 140–1
Kano Model, 120
Kolmorogov-Smirnov Test, 51
Krasnow Institute for Advanced Studies, 186
Kruskal-Wallis Test, 42
Kurtosis, 21–2, 31, 50, 52–3, 77

Lag Time, 133
Law of Large Numbers, 29, 77, 147
Lean, 136
Levene's Test, 53
Lewis Fry Richardson, 76, 98, 100, 108, 186
Liquidity Risk, 143
Listwise, 55
Lockheed Martin, 186
Log Transformation, 54
Lotka-Volterra, 97, 100

Macedonia, 125
MANCOVA, 42–4
Mann-Whitney U Test, 40
MANOVA, 42–4
Mardia's Statistic, 52
Marginal Cost, 126
Mass-Surveillance, 4, 105, 126
Mauchly's Sphericity Test, 53
Mean, 15–22, 24, 29, 31–2, 37, 39, 41, 45, 48–50, 52, 54–8, 77, 117, 133, 146, 165
Mean Squares, 41

Measure of Central Tendency, 15, 21
Median, 17, 19, 21–2
Middle East, 101, 103, 153–4
Minot Air Force Base, North Dakota, 47
MIRR, 163–4
Mode, 17, 22
Model, 2, 5, 10, 16–36, 38, 41, 43, 50, 55, 57, 61, 69, 72–81, 87–94, 97–8, 100, 103–4, 106–9, 113, 119–20, 122, 140–1, 144, 146–51, 167–79
Monte Carlo, 147
Motorola, 115
MRP, 134
MRS, 129
Multicollinearity, 56
Multiplier, 16, 30, 121, 159
Multivariate, 42, 52–3

Napoleon Bonaparte, 13
Nazi, 97–8, 104, 125
Neil Johnson, 108
Nominal, 13
Nonparametric, 40–2, 53
Nonprobability Sampling, 26–7
Normal Distribution, 20, 22, 31, 49–52, 73, 76–7, 95, 116, 147
Normality, 49–53
Normandy, 97–8
North Korea, 100
NPV, 162–3
Null Hypothesis, 8–9, 39, 44

OEF, 116, 130, 142
OIF, 116, 130, 142
Omaha, Nebraska, 26
Omega-Squared, 91
Operating Cycle, 132–7
Operating Risk, 142
Operation Bodyguard, 97
Operational Analytics, 5, 10, 16, 38, 57, 61, 111–75
Optimistic, 133
Oracle, 105

Ordinal, 13
Outlier, 49–50, 55, 77, 95
Outsource, 112, 131

Pairwise, 55
Pakistan, 100
Parabola, 34
Parametric, 38–42, 49, 53–4
Pareto Efficiency, 130–1
Partial Eta-Squared, 91
PDCA, 124
Pearson Chi-Squared Test, 43
Pearson's r, 86, 90
Personal Observation, 7, 11
PERT, 133
Pessimistic, 133
Poisson Distribution, 77–9
Polaris, 133
Population, 2, 15–16, 19–20, 24, 26–9, 35, 37, 46, 50, 67, 70, 77, 91, 95, 97–9, 147, 161, 184
Portfolio, 149–51, 160, 170
Power Transformation, 54
P–P Plot, 50
PPE, 164
Predictive Analytics, 3–4, 10, 57, 67, 69–109, 173
Prioritization Index, 122
PRISM, 94
Probability, 1, 8–9, 25–7, 29, 35–6, 39, 49–50, 56, 69–81, 84, 89, 95, 122, 140–50, 169–70, 175, 179
Probability Sampling, 26–7
Productive Efficiency, 126, 128, 131
Psychometrics, 5, 178
PV, 160–3
P-Value, 39–40, 44, 77

Q–Q Plot, 50–1
Quality, 3, 23, 36, 38, 88, 104, 115–24, 137, 154, 172
Quick Ratio, 165
Quota Sampling, 26

Random Sampling, 27, 29
Range, 12, 15–38, 44, 49, 55, 61, 65–7, 71, 74, 77, 92, 94–5, 104, 120, 122, 146–7, 153, 155, 187
Reciprocal Transformation, 54
Red Flag, 71, 94–6, 105
Reengineering, 136
Regression, 42, 45, 50, 58, 79–80, 83–93, 96, 101
Reinhart and Rogoff, 56
Reliability, 29, 47, 57, 63, 175
Riftland, 95
Risk, 5, 30, 55, 62, 107, 111, 113, 116, 122, 128, 132, 135, 139–51, 156–8, 164–74, 184, 186
Risk Assessment Matrix, 144
RMSSE, 45
ROA, 158–9, 162
ROE, 159
ROI, 157–8
Russia, 100

Sample Size, 28–9, 40, 56, 59, 67, 79
SCARE, 95
Scatterplot, 33–4, 86–7, 93
Shapiro-Wilk Test, 51
Sharpe Ratio, 150
Sigma, 16, 41, 162
SIPOC Analysis, 112
Six Sigma, 3, 115–18, 124
Skew, 21–2, 31, 37, 48–54, 77
Slope, 32–4, 50, 87–8
South Korea, 100
Southeast Asia, 101
Sphericity, 53
Squared Transformation, 54
Standard Error, 29, 37
Standard Deviation, 15, 20–4, 28–32, 45, 49–50, 77, 87, 116–17, 146, 150
Sterling Ratio, 151
Strategic Risk, 143
Stratified Sampling, 27
Student's T-Test, 31, 38–41, 44,

Sum of Squares, 41, 43, 91
Sun Tzu, 97, 170

Taguchi Loss Function, 118
Taiichi Ohno, 136
The Dupuy Institute, 186
Thomas Bayes, 80
Thomas Edison, 105
Thomas Schelling, 186
Toyota, 136
TPS, 136
TQM, 115
Tracy-Widom Distribution, 97
Transfer Pricing, 112, 131
Turnover, 135–6, 159, 165
Type I Error, 9
Type II Error, 9

UCLA, 97
UK (United Kingdom), 56
Uncertainty Risk, 143, 145
United States, 26, 70, 100, 126, 153–4, 185,
Univariate, 52–3
US Army, 115
US Army Corps of Engineers, 14, 184
US Chief of Staff of the Army's Strategic Studies Group, 154
US Civil War, 99
US Department of Defense, 3, 154, 185

US Department of Homeland Security, 94, 187
US National Security Agency, 94, 186–7
US Navy Special Projects Office, 133
USSR, 100

V.S. Subrahmanian, 186
V2 Ratio, 151
Validity, 55–8, 62–72, 140, 167
VaR, 146
Variance, 20–1, 24, 29–30, 41–3, 53, 56, 85–6, 91
Venn Diagram, 75–6
Vertex, 35
Volatility, 30, 140–1, 167, 173

WACC, 163–4
Wal-Mart, 96–134
W-ICEWS, 95
Wilcoxon Signed Rank Test, 41
Wilks' Lambda, 43
William Edward Deming, 124
William Gosset, 38–9
WWI, 125
WWII, 3, 56, 67, 96, 108, 115–16, 125, 137

X-Inefficiency, 131

Z-Score, 28, 49–50, 77, 83, 88

GPSR Compliance

The European Union's (EU) General Product Safety Regulation (GPSR) is a set of rules that requires consumer products to be safe and our obligations to ensure this.

If you have any concerns about our products, you can contact us on

ProductSafety@springernature.com

In case Publisher is established outside the EU, the EU authorized representative is:

Springer Nature Customer Service Center GmbH
Europaplatz 3
69115 Heidelberg, Germany

www.ingramcontent.com/pod-product-compliance
Lightning Source LLC
Chambersburg PA
CBHW071614100426
42873CB00004B/43